Michael Krystek
**Quantities and Units**

# Also of interest

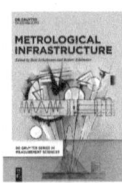

*Metrological Infrastructure*
Edited by Beat Jeckelmann, Robert Edelmaier, 2023
ISBN 978-3-11-071568-2, e-ISBN 978-3-11-071583-5
in
*De Gruyter Series in Measurement Sciences*
Edited by Klaus-Dieter Sommer, Thomas Fröhlich
ISSN 2510-2974, e-ISSN 2510-2982

*Geodesy*
Wolfgang Torge, Jürgen Müller, Roland Pail, 2023
ISBN 978-3-11-072329-8, e-ISBN 978-3-11-072330-4

*Solid State Physics*
Siegfried Hunklinger, Christian Enss, 2022
ISBN 978-3-11-066645-8, e-ISBN 978-3-11-066650-2

*Classical Mechanics*
Hiqmet Kamberaj, 2021
ISBN 978-3-11-075581-7, e-ISBN 978-3-11-075582-4

Michael Krystek

# Quantities and Units

The International System of Units

**DE GRUYTER**
OLDENBOURG

**Author**

**Dr. Michael P. Krystek** studied physics and mathematics at the Technical University of Berlin, received his PhD in physics and his habilitation in metrology. He was a Senior Scientist at PTB (Physikalisch-Technische Bundesanstalt) and chairman of IEC/TC 25 Quantities and Units until 2022. He was project leader for the last revision of the ISO/IEC 80000 series of standards in ISO/TC 12 Quantities and Units, PTB expert in the BIPM-CCU during the revision process of the International System of Units (SI), and expert for ISO and IEC in the JCGM working groups for VIM and GUM.

ISBN 978-3-11-134405-8
e-ISBN (PDF) 978-3-11-134411-9
e-ISBN (EPUB) 978-3-11-134416-4

**Library of Congress Control Number: 2023944392**

**Bibliographic information published by the Deutsche Nationalbibliothek**
The Deutsche Nationalbibliothek lists this publication in the Deutsche Nationalbibliografie; detailed bibliographic data are available on the Internet at http://dnb.dnb.de.

© 2023 Walter de Gruyter GmbH, Berlin/Boston
Cover image: Michael Krystek
Printing and binding: CPI books GmbH, Leck

www.degruyter.com

# Foreword

The community of metrologists present at the November 2018 General Conference of the Metre Convention (CGPM) resolved to fundamentally revise the International System of Units (SI). This revision came into effect on 20 May 2019. All seven base units, including the kilogram, the meter, the second, the ampere, and the kelvin, now have a unique definition by fixing the numerical values of "defining constants" — among them fundamental constants of nature such as Planck's constant and the speed of light, as well as the electron charge, Avogadro's constant, and Boltzmann's constant.

The over one hundred Member and Associated States of the Metre Convention voted unanimously in favour of this revision of the system of units, the most extensive since the Metre Convention itself came into force in 1875. As early as 1899, Max Planck had had the visionary insight to suggest such a definition in his famous article titled *"Über irreversible Strahlungsvorgänge"*. In this article, he postulated the Planck constant and thus founded the very field of quantum mechanics; the units defined in this way are "independent of special bodies or substances" and will "necessarily retain their meaning for all times and for all, even extraterrestrial and non-human cultures". The recent revision has made artefacts such as the prototype kilogram and the prototype meter obsolete — the unit system is now based on our modern theoretical understanding of nature.

Even for metrology, the science of measurement, and its applications, this step has meant a fundamental cultural change, even if units such as the meter and the second had long been defined according to this principle. Not surprisingly, the revision has therefore led to intense, fundamental, and controversial discussions among the members of the Metre Convention on basic concepts of metrology such as quantities, quantity values and units. This discourse continues to this day in a working group of the Consultative Committee for Units (CCU), as well as in a joint working group with other institutions of international quality infrastructure, which include the International Organization for Standardization (ISO). Another new working group of the CCU is pursuing a comprehensive debate on SI units that are merely an alias (i. e., a name or a designation) for the unit "one" or the symbol "1". These units include the radian, a unit formerly called a "supplementary unit" for the plane angle.

The author of this book has chosen just the right moment to prepare a comprehensive, internally consistent, modern, and clear presentation of the terminology of physical quantities, quantity values and units based on the revised SI. The book also includes a general overview of the basic elements of unit systems, an explanation of calculations involving units, a summary of internationally recognized notations and mathematical symbols, and an examination of common formula symbols. Be-

https://doi.org/10.1515/9783111344119-201

cause the revised SI is ultimately based on quantum theory as the foundation of modern physics, it follows that the treatise in this book concludes with a digression on nuclides, particles, and quantum states.

In analogy to the famous treatise *"Zählen und Messen, erkenntnistheoretisch betrachtet"* by Hermann von Helmholtz, the first president of the Physikalisch-Technische Reichsanstalt, this book also aims to provide orientation. The German Institute for Standardization (DIN) has also repeatedly developed fundamental concepts as well as terminology in metrology and contributed to the international discussion on this subject. Metrology (i. e., measurement) is becoming an increasingly important subject throughout the world: It is the basis of global trade, underpins the natural sciences and medicine, and will perhaps even become part of the social sciences and humanities in the future. Especially in view of this last reason, it is therefore crucially important that fundamental terminology and concepts of metrology be defined and established in an international consensus using current scientific methods. In this respect, this book will lend considerable impetus to the current international discourse.

The treatise contained in this book is timely for another reason: practical, applied, and legal metrology, like many other areas of life, is in a phase of upheaval, driven by digitalization. Metrology is not only a science — it also represents a service used by people, the economy, and society. The process of digitalization continues apace throughout the world: In the metrology of the future, calibration certificates, permits, accreditations and many other documents will be created completely digitally. In the long term, it will be necessary not only for calibration certificates to be machine-readable, but even for machines to interpret and act on them without human assistance. For this to happen, logically consistent terminology, semantics, and ontology must be available in addition to internationally agreed-upon metadata formats for measurement data. The revision of the system of units will give the science of metrology a firm scientific foundation far into the future; this, in turn, will allow the community of metrologists, together with international quality infrastructure organizations, to tackle this new, major challenge.

At the CGPM in November 2022, a Draft Resolution will be presented to establish the Metrology and Digitalization Forum, a platform that will take on this very task. For all these reasons, I would like to wish the author and the treatise the attention they deserve during this international discussion on the future of metrology.

Braunschweig, January 2022
Prof. Dr. Dr. h. c. Joachim H. Ullrich
President of the *Physikalisch-Technische Bundesanstalt* (PTB)
Vice President of the *Comité International des Poids et Mesures* (CIPM)
President of the *Comité consultatif des unités* (CCU) of the CIPM

# Preface

In 1875, the Metre Convention was signed by 17 countries in Paris, a treaty that led to an international standardization of physical units for the first time in the more than five thousand years history of counting and measuring. The Metre Convention, which elevated the metric system created after the French Revolution of 1789 from a national to an international system, gave rise to the International System of Units (SI) in 1960, which is now used worldwide.

Initially, in 1889, only the units for mass and length were defined by measuring standards made of a platinum-iridium alloy, as was customary at this time. Over the years that have passed since then, the original system of units has undergone considerable changes. It has been gradually expanded and modernized, and there have also been several redefinitions, which — largely unnoticed by the majority of physicists — led to a progressive abstraction of the term *physical unit*. The reader can gain information about these developments from the first chapter.

The remaining chapters of this book have more the character of a reference and are intended to inform as comprehensively as possible, but not in depth, about the International System of Units (SI) as it appears after the revision in 2019.

The book is intended for all readers who deal with physical quantities and their measurement, i. e. natural scientists, engineers, measurement technicians, but also teachers, university professors, and students of these scientific disciplines.

The first three chapters of the book also provide mathematicians who wish or have to deal more with the theoretical aspects of metrology with sufficient information to familiarize themselves with the terminology of metrology and thus facilitate the exchange of experiences with other scientific disciplines.

At this point I would like to thank Dr. Franz Josef Drexler for the many fruitful discussions and for many years of friendly cooperation. The designation $|Q|$ for the value of a quantity $Q$ is his idea, which I have gratefully adopted.

I also would like to express my sincere thanks to Prof. Joachim Ullrich for the many years of close and trustful cooperation and for his willingness to write the foreword for this book.

Finally, I would like to take this opportunity to thank the De Gruyter publishing team for their efficient collaboration in making this book possible. My special thanks go to Mrs. Karin Sora for her spontaneous and kind decision to include this book in the de Gruyter publishing program and Mrs. Kristin Berber-Nerlinger for supporting her decision.

Berlin, January 2022
Michael Krystek

https://doi.org/10.1515/9783111344119-202

# Contents

# 1 Introduction

ἀριθμῷ δέ τε πάντ᾽ ἐπέοικεν

καὶ πάντα γα μὰν τὰ γιγνωσκόμενα ἀριθμὸν
ἔχοντι : οὐ γὰρ ὀτιῶν τε οὐδὲν οὔτε νοηθῆμεν
οὔτε γνωσθῆμεν ἄνευ τούτω

*Everything corresponds to number.*

*And indeed all things that are known have number.*
*For it is not possible that anything whatsoever be*
*understood or known without this.*

PHILOLAUS OF CROTON

**1.** Metrology deals with *counting* and *measuring*. There are only few things that affect people so directly and universally. Similar to the use of language by a people, the everyday use of their numbers and measures happens intuitively and without active thought.

We can only conjecture about the origin of numbers and measures because there is a lack of sufficient historical sources. However, we can assume that the idea of counting and measuring can be traced back to elementary mental processes and needs of humans.

Numbers probably emerged very early in human history through an abstraction of counting. Notched bones from the Stone Age can be interpreted as the beginnings of the abstraction process that ultimately led to the emergence of the concept of number. The connection between the original terms *number* and *enumeration* is still recognizable in most languages.

Numbers were originally considered to be composed of units. This view can still be found in the first two definitions in book VII of EUCLID's *Elements*. Numbers in Euclid's time were exclusively today's *natural numbers*. Their unit, the number one, was not initially regarded as a number because of its special properties.

Counting is based on the idea that the objects of the surrounding world (animals, trees, stones, etc.), which are to be understood as indivisible entities, can be grouped and arranged either actually or at least mentally. The question of "How many?" we answer with a number obtained by counting. This number, the *number of entities*, is always a *number of something*.

The idea of measurement probably also arose very early. It is based on the perception that not all things in the surrounding world are composed of separate

https://doi.org/10.1515/9783111344119-001

entities. It will not have escaped the notice of early humans that e. g. a certain quantity of water, which was essential for them, could not be determined by counting, because it could be divided arbitrarily and therefore each of its parts could serve as a unit. In order to be able to communicate with other people, it was therefore also necessary to name the underlying partial quantity (e. g. "a handful") that served as a unit. To the question "How much?" we answer with a number obtained by measuring. But this number only makes sense in connection with a unit. This unit, however, can be chosen arbitrarily.

**2.** The development of numbers and measures from the early days of mankind to the emergence of the first advanced civilizations about 5000 years ago lies in the darkness of history. But it seems to have been similar everywhere.

In Egypt and Mesopotamia, there was already a fully developed system of units thousands of years ago with mandatory units of length and weight that were represented by material measures. Since it was already possible to calculate with fractions and a number system existed for the representation of arbitrary large numbers, it was possible to deal with parts and multiples of these units. The measurement of lengths and weights could be carried out by a direct comparison with their respective material measures.

Time measurement had to be done in a different way. In order to determine calendar dates and time, the observation of the celestial bodies was used, especially the sun and the moon. The duration of the earth's rotation served as the basis for the unit of time, the *day*, which was initially divided into twelve double hours. This division had its origin in the star mythology of the Mesopotamians. Then, in the 6$^{\text{th}}$ century BC, Babylonian astronomers introduced the division of the day into 24 hours. This division of the day and the use of the *second* as a unit of time duration, which is still common today, is more than 2500 years old.

**3.** The measuring techniques, the calculation methods and the astronomical knowledge of the advanced ancient cultures, which were already astonishingly well-developed for their time, were adopted from Mesopotamia and Egypt almost unchanged by the Ancient Greeks in the 5$^{\text{th}}$ century BC. The Greek astronomers, deviating from the decimal system otherwise used in Ancient Greece, even used the sexagesimal numeral system[1] common among the Babylonians.

---

1 The sexagesimal numeral system has the number 60 as its base, while the decimal system has the number 10 as its base. The sexagesimal numeral system is still used today in a modified form to express the values of time, angles, and geographic coordinates.

The Greeks further developed the knowledge and skills inherited from the other cultures. Around the 4th century BC, the technique (ancient Greek τέχνη, i. e. *skill*) of measuring gave rise to the beginnings of metrology (ancient Greek μετρεῖν, i. e. *to measure*, and λόγος, i. e. *right insight* or figuratively *science*) and the practical calculation methods of the Egyptians and Babylonians led to the beginnings of mathematics with definitions, axioms and proofs. The traces of this development can still be found today in the writings of ARISTOTLE and in the *Elements* of EUCLID.

Even in pre-scientific times, people were certainly aware that there exist characteristics of the things in their surroundings that are quantitative (i. e. capable of *more or less*) and that these quantitative characteristics can be of different natures, namely either countable or measurable. But it was ARISTOTLE who specified these facts of experience in his *Theory of Categories* and *Metaphysics*. He distinguishes between *quantity* (ancient Greek ποσόν) as a generic term and *magnitude* (ancient Greek μέγεθος) on the one hand and *multitude* (ancient Greek πλῆθος ὡρισμένον, i. e. *determined number*[2]) on the other hand as specific terms. Of quantity he says that it is partly discrete (e. g. number), partly continuous (e. g. line and time). Of multitude, he says that it is countable and of magnitude that it is measurable. Comparability, additivity and the possibility of order are for ARISTOTLE also already essential properties of any quantity.

Remarkably, there existed already an *ontology* of counting and measuring in the 4th century BC. The next step was then the still missing elaboration of definitions and axioms. This was done by the ancient Greek in the following century and is summarized in the *Elements* of EUCLID. Of these, in addition to the first four books on plane geometry, especially Book V, which contains the theory of proportions established by EUDOXUS OF CNIDUS, Book X, which traces its content back to THEAETETUS OF ATHENS and deals with the geometry of incommensurable quantities, as well as the axioms (ancient Greek κοιναί ἔννοιαι, i. e. *general meanings*) in Book I were important for the development of metrology into a science.

**4.** It was already known 5000 years ago that every measurement is based on a comparison of the magnitude to be measured with a magnitude of the same kind determined by agreement, which serves as a unit. For the advanced ancient cultures, the measurement of lengths and weights[3] was particularly important. The units of these magnitudes were, as already mentioned, represented by material measures. This was still unchanged when the Metre Convention was signed in 1875.

---

2 The physicist HELMHOLTZ still used this term in his writings to express that he was referring to a variety of specific things, not abstract mathematical numbers.

3 The term "mass" was still unknown in antiquity.

At the first session of the *Conférence Générale des Poids et Mesures*[4] (CGPM) in 1889, the following two resolutions, among others, were adopted:

*1. The Prototype of the metre chosen by the CIPM.[5] This prototype, at the temperature of melting ice, shall henceforth represent the metric unit of length.*
*2. The Prototype of the kilogram adopted by the CIPM. This prototype shall henceforth be considered as the unit of mass.*

The first system of units established by *international* agreement (usually called the metric system) was thus still based on an embodiment of the units of length and mass. Even more, by these resolutions it was stipulated that the prototypes themselves should be the units of length and mass respectively. This is utterly consistent with the spirit of ARISTOTLE, who regarded line, surface and body itself as magnitudes, a view that was superseded by the 19th century. Apparently, however, despite all the advances in physics and mathematics that had led to a substantial improvement in measurement technology, the view that the prototypes themselves were the units was still very much alive.

The definitions of the CGPM from 1889 were only corrected decades later, namely in 1901 by the resolution

*The kilogram is the unit of mass; it is equal to the mass of the international prototype of the kilogram;*

adopted at the 3rd meeting of the CGPM and in 1927 by the resolution

*The unit of length is the metre, defined by the distance, at 0°, between the axes of the two central lines marked on the bar of platinum-iridium kept at the Bureau International des Poids et Mesures and declared Prototype of the metre by the 1st Conférence Générale des Poids et Mesures, ...*

adopted at the 7th meeting of the CGPM. According to the changed definitions of the units of length and mass, these units were no longer the material measures themselves, but the length and the mass of the respective prototype, as corresponded to the common understanding of physical magnitudes at the beginning of the 20th century. An important change had occurred. Quantities were no longer concrete objects, but the *quantitative characteristics* of these objects. This process of abstraction had

---

4  The name component *Poids et Mesures* means weights and (length) measures
5  CIPM is the abbreviation for *Comité international des poids et mesures*

become necessary because in physics there were now quantities, such as energy, that could not be embodied but were of a purely abstract nature.

In addition to the traditional *basic quantities*[6] of length, time and mass, *derived quantities* had to be introduced now. The measurement of these quantities is no longer possible by comparison with material measures.

**5.** The representation of the unit of length by material measures no longer met the requirements of precision metrology at the end of the 19th century. Ideally, units should be invariable and should be representable at any location and at any time. However, it was also clear that this ideal could only be realized by an atomic phenomenon. This had become possible through the development of physics and metrology at the beginning of the 20th century. By the resolution

> *The metre is the length equal to 1 650 763.73 wavelengths in vacuum of the radiation corresponding to the transition between the levels $2p_{10}$ and $5d_5$ of the krypton 86 atom.*

during the 11th CGPM in 1960, the unit of length was redefined. The numerical value used in the definition was chosen in such a way that, within the measurement uncertainty that was feasible at that time, there was no noticeable change compared to the previously valid unit of length; so the name could be retained.

The metre was the first unit whose definition was based on a physical phenomenon considered to be invariable. The unit of length could now be realized at any location and at any time, independent of material measures and measurement procedures, provided that the experimental facilities were available.

The procedure used in the redefinition of the metre – and also in later revisions of the SI – presupposes the definition of the term *quantity value*. It can be described as follows: The quantity values of the quantities to be redefined are determined by measurement and represented by using the units of the system of quantities existing at this very time. The measured numerical values of the quantities (without their measurement uncertainty) are then considered exact by agreement, which eventually redefines the units.

**6.** In the first meeting of the CGPM, only the units for the measurement of mass and length were determined, but there was no determination of the unit of time. In

---

6 From a mathematical point of view, these quantities are not necessarily identical with the *base quantities* of a system of units. Base quantities can in principle be arbitrarily chosen, but they must fulfil mathematical conditions with regard to their transformation behaviour.

practice, however, as had been the practice since the Babylonians, the unit of time, the *day* — determined by the rotation of the Earth and defined as the duration of the mean solar day — was used, which was divided into hours, minutes and seconds by means of clocks, in the manner that had been customary for millennia.

In 1935, using quartz clocks at the Physikalisch-Technische Reichsanstalt (PTR) in Berlin, it was demonstrated that, in addition to the gradual decrease in the Earth's rotation period due to tidal friction[7] known since the beginning of the 20th century, there are also unpredictable variations in its rotation. On the other hand, by the end of the 19th century, computational methods were already available to calculate the orbits of celestial bodies very precisely from their observed positions. This made it possible to trace the duration of a second back to the much more regular annual movement of the Earth around the Sun. In 1956 the *International Astronomical Union (IAU)* therefore made the proposal to define the unit of time as a certain fraction of the tropical year.[8] Following this proposal, the CGPM decided in 1960 at its 11th meeting by resolution 9 the following definition of the SI unit of time:[9]

> *The second is the fraction 1/31 556 925.9747 of the tropical year for 1900 January 0 at 12 hours ephemeris time.*

This so-called "ephemeris second" (ancient Greek ἐφήμερος, i. e. *for one day*) can only be as accurate as the formula established to predict the Earth's orbit. Later verification showed that it was about 30 ns shorter than the *Universal Time* second valid today. This deviation did not play a significant role in the determination of the civil time, because it corresponds to about one second per year.

The *Ephemeris Time* was a *theoretically* defined *ideal* time running completely uniformly on the basis of the known laws of nature. Thus the ephemeris second was a different unit of time than the second of the civil determination of time and had only the name in common with it. However, the constants in the formula for the calculation of the Ephemeris Time had been chosen in such a way that the *quantity value* 1 s of the unit of time remained unchanged and on 0 January 1900 (i. e. 1899 December 31) at 12:00 o'clock the ephemeris time coincided as closely as possible with universal time UT.

The realization of the ephemeris second, which was no longer related to the Earth's rotation, had to be done by observing the Moon, because time intervals can

---

7 This shortens the year by about one day in 10 million years.

8 The tropical year is the period of time between two successive passages of the "mean sun" through the "mean vernal equinox".

9 In 1960, the CGPM also decided that the system of units would henceforth be called *Système International d'Unités (SI)* (i. e. International System of Units).

be determined only very imprecisely from the Earth's orbit. The ephemeris time was therefore ultimately based on the complex theory of the Earth-Moon movement. A uniformly moving time could therefore only be assumed in theory. In practice quartz clocks were already used in the thirties of the 20th century to represent standard time and as frequency standards.

7.  The problems of astronomical time determination and the fact that since the thirties of the 20th century, quartz clocks with a relative uncertainty of $10^{-9}$ and since 1955 already caesium atomic clocks existed for scientific applications,[10] makes the introduction of ephemeris time by the CGPM in 1960 appear as a curiosity. Then, in 1967, the resolution

> *The second is the duration of 9 192 631 770 periods of the radiation corresponding to the transition between the two hyperfine levels of the ground state of the caesium 133 atom;*

adopted at the 13th CGPM, revised the unit of time again.

The basis of this definition was the number of oscillations of the underlying microwave radiation of the caesium atom determined with very accurate quartz clocks, which corresponded to the ephemeris second with a stated uncertainty of ±20 periods. The unit of time had thus finally been reduced to the counting of periods of a uniform periodic process in accordance with the theoretical ideal.

8.  The speed of light in vacuum is considered a fundamental constant of nature, i. e. as independent of location and time. The quantity value of this constant had to be determined by measurements. Since both the unit of length and the unit of time in the SI had been based on atomic phenomena since 1967, these measurements could be carried out in a well-reproducible manner with a very small measurement uncertainty. Therefore, in 1975, with Resolution 2 of the 15th CGPM, it was decided that the quantity value $c = 299\,792\,458$ m/s of the speed of light in vacuum, known at that time, should be regarded as exact by agreement.

The decision of the 15th CGPM led to a dilemma, because the unit of length, the unit of time and the quantity value of the speed of light based on these units could

---

10 Resolution 10 of the 11th CGPM in 1960 even explicitly states that *"…appreciating the experimental results obtained by competent laboratories in recent years, which prove that a time interval standard based on a transition between two energy levels of an atom or molecule can be realized and reproduced with very high accuracy,"*.

not be exact at the same time.[11] This problem was only solved by Resolution 1 of the 17th CGPM in 1983. With the new definition of the unit of time

*The metre is the length of the path travelled by light in vacuum during a time interval of 1/299 792 458 of a second*

the unit of length now became a unit derived from the unit of time and the quantity value of the speed of light.

**9.** In the redefinition of the unit length, the connection between the quantity values of the constants of nature and the units of a system of units used to represent them became clearly visible. The units are required to be ideally independent of time and location. However, this is also an essential property of the constants of nature.

The presupposed constancy of the constants of nature implies a constancy of their quantity values with respect to an arbitrarily chosen system of units. This results in two principally different possibilities, namely

1. the establishment of the units of a system of units by agreement and based thereon the determination of the numerical values of the constants of nature in relation to this system of units,
2. the fixing of any numerical values of the constants of nature with respect to a system of units by agreement and the definition of the units of that system on the basis of the numerical values fixed in this way.

The first option was used in the past, the second for the first time in 1983 when the unit of length was redefined, and finally in 2019 when the SI was revised, although not arbitrary numerical values were chosen for the constants of nature, but the values already known from past measurements, so that the revision of the SI has not led to noticeable changes of quantity values in science and technology.

**10.** By decision of the 10th CGPM in 1954, the ampere was introduced as a SI base unit. The definition of the ampere was based on the force exerted on each other by two infinitely long parallel conductors of negligible cross-section when an electric current flows through them. This definition, which was valid until the revision of the SI in 2019, was unusable for an application in electrical engineering. Therefore, in 1988, the CIPM decided to adopt a more practical definition based on quantum

---

**11** The measurement uncertainty could not simply disappear, but had to be transferred to the unit of length or the unit of time. But this is impossible, since units are *by definition* exact.

mechanics. In doing so, use was made of the second possibility mentioned above, namely defining units via natural constants.

The decision of the CIPM from 1988 was eventually implemented in practice on 1.1.1990. The units of the voltage and resistance were defined by specifying the quantity values of the JOSEPHSON constant and of the VON KLITZING constant, respectively. The unit of electric current can then be calculated via Ohm's law.

Both the JOSEPHSON constant and the von KLITZING constant depend only on the two constants of nature $e$ (elementary charge) and $h$ (Planck constant). Because the former quantity values of $h$ and $e$ were no longer valid since the revision of the SI in 2019, the definitions of the units of voltage and electrical resistance had to be amended accordingly. After these only minor corrections, electrical measurement technology is now again based on the SI after almost thirty years.

**11.**    After the unit of mass was also redefined by the revision of the SI in 2019, the last material measure in the SI was also eliminated and all SI units are now based on constants of nature or otherwise determined constants, which are collectively called "defining constants". This has now created a system of units based on modern knowledge of the natural sciences, which will not undergo any major changes in the near future. Only for the unit of time a change is to be expected in the not too far future, whereby, however, the principle will not change, but a different atomic transition with a higher frequency in the optical spectral range will be used as a basis. With this "optical atomic clock", a considerably smaller uncertainty of time measurement can then be achieved. A further increase in measurement precision is possible if excited levels within the atomic nucleus are used instead of an electron transition in the atomic shell. As was experimentally demonstrated in September 2019, the construction of "nuclear clocks" using the nuclide $^{229}$Th as the currently only candidate is already within the realm of possibility.

**12.**    The development of mathematics during the two millennia that had passed since antiquity had led to a change in the concept of "number", which had been considerably expanded. Mathematics had also dealt with the theoretical foundations of metrology.

The introduction of the real numbers eliminated the problem of incommensurable quantities, so that in metrology today we can express all quantity ratios by numbers (in the sense of the now expanded concept of numbers). However, this must not be understood to mean that real numbers (i. e. ultimately the continuum) are *measurable*. The result of any measurement, no matter how precise, is and always remains a rational number. If we express the ratio of the length of the diagonal of a square to the length of its sides by the symbol $\sqrt{2}$, then this is an idealization that is not accessible to measurement. We cannot verify the Pythagorean theorem by measurement.

The fact that measurement results can only be expressed by rational numbers does not impose any restriction in practice, because every real number can be approximated by a rational number as precisely as desired. Moreover, the ever-present measurement uncertainty does in any case not allow us to determine exactly the equality of two quantities of the same kind by means of a measurement.

Another advantage of the introduction of real numbers in metrology is that the ratio[12] of two quantities of the same kind can be represented by a number, i. e. for each quantity a *quantity value* can always be stated as a product of a number and a unit, where the number has to be determined by measurement. This process of abstraction[13] is the mathematical basis of theoretical physics, because it ensures that physical hypotheses (the laws of physics) can be tested by measurements without having to renounce the advantages of analysis.

Strictly speaking, the formula symbols of physical equations do not denote quantities, but quantity values. One hundred years ago, this insight was the basis for the introduction of calculation (computation) with quantity values by JULIUS WALLOT. However, as was customary in physics at that time, he did not make a distinction between the terms *quantity* and *quantity value*, so that today this technique is often wrongly referred to as "quantity calculus"[14] although in fact it is not the quantities themselves but the quantity values that are subject to computation.

**13.** In recent decades, the term "measurement technology" has often been replaced by the term "metrology", in order to emphasize the scientific aspect of measurement. However, a science is expected to strive for clear and unambiguous terminology.

So far, however, there is no agreement on the definitions of fundamental terms of metrology, such as "measurement", "quantity", "magnitude", "quantity value", and "unit". It is even discussed whether the terms "quantity" and "magnitude" need to be distinguished or whether the term "quantity value" is necessary at all. In order to reach a decision on this and other related questions and to arrive at a logically consistent terminology, it would be worthwhile to take a look at the definitions commonly used in mathematics.

---

**12** Ratios are *relations* and must not be confused with *quotients*.

**13** Mathematically, this is a bijective mapping to the real numbers.

**14** By the way, Julius Wallot never spoke of a "quantity calculus", but only of calculating (computing) with quantities and units. Presumably he was aware that a calculus is something else, namely a formal system of rules with which further statements can be derived from given statements. Calculi usually exist in the formal sciences, such as logic, mathematics and computer science.

Definitions of terms using only a natural language will no longer be sufficient in the future. If we want to use computers to check the logical consistency of a system of terms (this is already possible today), then terms must be defined in a formal language. This would have the additional advantage to prevent misunderstandings caused by translations from one natural language to another. In case of doubt, the formal definition would then always apply.

# 2 Physical quantities

## 2.1 Quantity, unit, quantity value

A quantity is either a multitude, if it is discrete, or a magnitude,[15] if it is continuous (as already mentioned in the introduction, this fact was familiar even to the ancient Greeks and stated by ARISTOTLE in his Categories). A multitude can be counted, e. g. a countable set of entities, and a magnitude can be measured. Physical quantities, such as length or mass, are thus magnitudes.

There are two different meanings of the term *quantity*. One refers to an abstract metrological concept (e. g. duration, length, mass, temperature, etc.), the other to a concrete realization of this concept by a characteristic of a certain physical object or phenomenon (e. g. the period of oscillation of a pendulum, the length of a copper wire, the propagation velocity of water waves, the mass of the proton, etc.). Both meanings are based on the term *quantitative characteristic*.

A characteristic is a recognizable attribute that makes it possible to distinguish one object or abstract context from others. Characteristics also play a special role in the sciences for the classification of objects and phenomena according to certain criteria (classification scheme, taxonomy), whereby they are arranged in categories or classes. In the natural sciences, we are particularly interested in *quantitative* characteristics. Their characteristic values are countable or form a continuum.

Characteristic values of a quantitative characteristic can always be compared with each other, i. e. in principle it can always be determined whether two characteristic values are the same or whether one is larger (or smaller) than the other. The property of total comparability is shared by the characteristic values of a quantitative characteristic with the numbers, but without being numbers for that reason.

On the basis of the term quantitative characteristic, we can now define the term *quantity* as

**Quantity.** *A quantitative characteristic is called quantity if there exists a ratio[16] for each pair of its characteristic values to which a real number is assigned.*

---

15 A hundred years ago, it was still common in metrology to make a distinction between quantities and magnitudes. Today these terms are unfortunately often used as if they have the same meaning. In fact, any magnitude is a quantity, but not vice versa.

16 Note that a ratio is *not* a quotient, which is the result of the division of two numbers, but rather a relation.

https://doi.org/10.1515/9783111344119-002

This definition of the term *quantity* initially entails a restriction to scalars. But there are quantities that are vectors or tensors. In these cases, however, their components can be treated as scalar quantities. We can therefore restrict the following reasoning to scalar quantities.

Real numbers can be assigned to the characteristic values of a quantity in a one-to-one correspondence by assigning the number one to exactly one characteristic value. All others are then multiples or fractions of this special characteristic value. We therefore define the term *unit* as

> **Unit.** *The characteristic value of a quantity to which the number one is assigned by agreement is called the unit of the quantity.*

and the term *quantity value* as

> **Quantity value.** *The representation of a characteristic value of a quantity as a product of a number and a unit is called quantity value, i. e.*
>
> $$|Q| = \{Q\}[Q]$$
>
> *applies, where $|Q|$ denotes the quantity value of the quantity $Q$ and $[Q]$ its unit. The number $\{Q\}$ is called the numerical value of the quantity.*

The unit of a quantity is denoted by its agreed unit symbol and the preceding numeral for the number one, the digit 1. If the quantity is a pure number, then its unit is the number one, and it is only represented by the digit 1. In a product or quotient of units or quantity values, the digit 1 may be omitted, but care must always be taken that unit symbols are not allowed to stand alone.

| *Examples:* | | |
|---|---|---|
| | 3.6 N | (Quantity value of a force in the SI) |
| | 1 N | (Unit of force in the SI) |
| | $300\,\text{m}^2$ | (Quantity value of an area in the SI) |
| | $1\,\text{m}^2$ | (Unit of area in the SI) |
| | $120\,\text{N/cm}^2$ | (Quantity value of a pressure in the SI) |
| | $1\,\text{N/m}^2$ | (Unit of pressure in the SI) |

## 2.2 Systems of quantities

A system of quantities consists of a set of quantities and relations between them. Some of these quantities, which are considered to be pairwise independent of each other and of all other quantities, form the set of base quantities for the entire system. All other quantities of the system of quantities are then derived quantities, defined in terms of the base quantities and expressed algebraically by products of powers of the base quantities.

To each quantity a dimension with the same name is uniquely assigned in a one-to-one correspondence. The representation of a quantity as a product of powers of the base quantities is therefore usually called the dimensional product (or physical dimension) of the quantity with respect to the chosen base quantities or base dimensions. The integers used as exponents of the base quantities or base dimensions are called dimensional exponents.

In the *International System of Quantities (ISQ)* there are seven base quantities and accordingly seven base dimensions, each sharing the same name. The base dimensions are denoted by upright, sans-serif capital letters (see table 2.1).

**Tab. 2.1:** Names and symbols of the base dimensions in the ISQ.

| Name of quantity and dimension | Symbol of dimension |
| --- | --- |
| (time) duration [†] | T |
| length | L |
| mass | M |
| electric current | I |
| thermodynamic temperature | $\Theta$ |
| amount of substance | N |
| luminous intensity | J |

[†] Time is not strictly a quantity (nor is physical space). In physics, in fact, only (time) duration is used, i. e. a time interval, which is a quantity comparable to the length of a distance.

In the ISQ, dimensional products are written in the form

$$\dim Q = \mathsf{T}^{\alpha}\mathsf{L}^{\beta}\mathsf{M}^{\gamma}\mathsf{I}^{\delta}\Theta^{\varepsilon}\mathsf{N}^{\zeta}\mathsf{J}^{\eta},$$

where $\dim Q$ denotes the physical dimension of the quantity $Q$ and $\alpha$, $\beta$, $\gamma$, $\delta$, $\varepsilon$, $\zeta$, $\eta$ denote the integer dimensional exponents.

Quantities whose dimensional exponents are all equal to zero are called quantities of dimension number. These quantities (e. g. angular measure, refractive index, etc.) are numbers. Their quantity values are expressed only by numerical values. Their unit, the number one, is usually omitted, but can be expressed by a supplementary unit if necessary (see section 3.4).

The dimension number cannot be a base dimension in any system of quantities (for details see [1]). It is denoted by the letter Z. In the ISQ therefore the following applies

$$Z := T^0 L^0 M^0 I^0 \Theta^0 N^0 J^0 \,.$$

Thus the relation $\dim Q = Z$ expresses the fact that the quantity $Q$ is a number and consequently for its unit $[Q] = 1$ is valid.

A system of units can be unambiguously assigned to each system of quantities by defining a base unit for each base quantity and expressing all other units of the system of units by products of powers of the base units so that the relations existing between the quantities correspond to those between the units. The dimensional exponents of the base units are identical to those of the corresponding base quantities and base dimensions.

## 2.3 Symbols for physical quantities

Symbols for physical quantities shall consist of single capital or small italic letters of the Latin or Greek alphabet with or without supplementary symbols (subscripts or superscripts, accents, underlining or overlining, etc.). The only exception to this rule are the two-letter symbols used to represent characteristic numbers (see section 6.14). If such a two-letter symbol appears as a factor in a product, it shall be separated from the other symbols by a dot (multiplication sign), a small space or a bracket. It is treated as a single symbol and can be raised to a positive or negative power without the use of a bracket.

Abbreviations (i. e. abbreviated forms of names or expressions, such as e. g. pdf for probability density function) may be used in the text, provided they have been previously introduced, but they shall not be used in physical equations. Abbreviations in the text shall be written in standard Latin script.

Symbols for physical quantities and numerical variables shall always be written in an italic font. This also applies to indices denoting quantities or numbers, while explanatory indices shall be written in standard (upright) script. Numbers are always written upright.

*Examples:* $C_p$                       ($p$: pressure)

$E_{\mathrm{kin}}$                (kin: kinetic)

$\check{x} := x_{\mathrm{min}}$         (min: minimal)

$T_i^k$                      ($i, k$: tensor indices)

$$\sum_{n=1}^{3} a_n x^n = a_1 x + a_2 x^2 + a_3 x^3 \qquad (n\text{: an index})$$

It is appropriate to use unique fonts for symbols in order to distinguish between the components of a vector or tensor and the vector or tensor as a mathematical object in its own right, or to avoid the use of indices. The following standard conventions shall be followed if the respective fonts are available:

(a) vectors shall be written in bold italics, e. g. $\boldsymbol{v}$.
(b) tensors shall preferably be written in bold italics without serifs, e. g. $T$.

For vectors, normal italics with one arrow can also be used, e. g. $\vec{v}$. For tensors of rank two, normal italics with two arrows or with a double arrow may alternatively be used, e. g.

$$\vec{\vec{T}} \quad \text{or} \quad \overset{\leftrightarrow}{T}.$$

The application of this notation to tensors of higher rank is awkward; in these cases the index notation shall be used throughout for tensors and vectors.

*Examples:* $\quad v_i \quad T_{ij} \quad R^{ij}_{kl} \quad \delta^k_i$.

## 2.4 Calculating with quantity values

When calculating with quantity values, all rules of mathematics apply in the first instance. However, there are certain restrictions that we have to deal with here, because there is a danger that ignoring them will lead to nonsensical results.

Quantity values can be added and subtracted, provided they are of the same kind. However, it is important to realize that quantities of the same kind are not necessarily identical with quantities of the same unit. The reason for this is the fact that in practical unit systems such as the SI, the same unit may be assigned to different quantities.

It is e. g. nonsensical to add the quantity values of work and torque, even though they have the same SI unit, because they are not of the same kind. They are not only physically different, but also mathematically; work is defined as a scalar product of two vectors and torque as a vector product.

Quantity values can be multiplied and divided. This also includes exponentiation with integer exponents. Multiplication and division are most easily done by treating numerical values and units separately. When calculating with units, all units should first be converted to SI units, if necessary, and the prefix symbols (see section 3.5) of units shall be converted to numbers.

*Example:*
$$36\,\text{km/h} = (36 \times 10^3\,\text{m})/(3600\,\text{s}) = (36 \times 10^3/3600)\,\text{m/s} = 10\,\text{m/s}$$

# 3 Units

## 3.1 Systems of units

A system of units is based on a set of base units that are assigned to the respective base quantities by convention. The number of base units of a system of units is equal to the number of the corresponding base quantities of the underlying system of quantities.

The numbers are also always part of every system of quantities. Their unit is the number one, symbol 1. The number one is a special unit because it is the neutral element of multiplication; every unit remains unchanged when multiplied by the number one. Because of this special property, it cannot be a base unit in any system of units (see [1] for details).

All units for derived quantities are expressed as products of powers of the base units, fully analogous to the corresponding expressions in the system of quantities. The system of units and its units are called coherent if the derived units are expressed by relations in such a way that all numerical factors are equal to one.

Derived units and their symbols are expressed by the base units using the mathematical signs for multiplication and division. Some derived units have been given special names and symbols—mostly for historical reasons—which can also be used to form names and symbols of other derived units.

Since the main purpose of a system of units is to provide a basis for converting the numerical values of physical quantities whenever the units are changed, and since quantities of dimension number [17] are invariant under such a conversion, there is no need to include quantities such as the angular measures of the plane angle and the solid angle in the category of basic quantities. For mathematical reasons, this is not even possible, since the unit of all quantities of dimension number is the number one. However, as already stated above, the number one cannot be a base unit in any system of units because of its special property.

## 3.2 The international system of units (SI)

The name *Système International d'Unités* (International System of Units) with the international abbreviation SI was adopted by the *Conférence Générale des Poids et Mesures* (CGPM) in 1960. The SI has been revised from time to time thereafter to

---

[17] Quantities of dimension number are often still incorrectly called *dimensionless* quantities. However, this term is deprecated since several years.

https://doi.org/10.1515/9783111344119-003

meet user requirements and advances in science and technology. The latest revision of the SI was adopted by the 26th CGPM (2018) and came in force on 20th of May 2019. It is documented in the 9th edition of the SI brochure [18] [2].

The SI is a consistent system of units for all fields of life, including international trade, production, safety, health, environmental protection and the scientific foundations on which all these fields are based. The system of quantities underlying the SI and the associated defining equations are based on the currently accepted description of nature (the so-called laws of nature) and are familiar to all scientists, engineers and technicians.

The definition of the SI units of the revised SI of 2019 is done by means of seven so-called *defining constants*, which are only to some extent natural constants and whose numerical values expressed by the existing SI units are now regarded as invariant.[19] These seven constants are the basis of the definition of the base units of the revised SI, because if they are seen as a set of base units, then the base units of the SI can be obtained from them simply by performing a base transformation. The *defining constants* were selected on the basis of the advances in precision metrology during the past decades and in consideration of the practical application of the existing definitions of the SI units.

For the realization of the definitions of the SI units, experimental methods called *mises en pratique* have been elaborated by the advisory committees of the CIPM, the descriptions of which are available on the BIPM website. They will be revised in the future whenever new experimental methods are developed on the basis of new scientific knowledge.

For the realization of the unit of time, intensive work is currently in progress in order to reduce the measurement uncertainty by two to three orders of magnitude by means of *optical clocks*. It is therefore to be expected that the definition of the unit of time in the SI will change again in a not too far future.

Individual countries have laid down rules for the use of units by national legislation either for general use or for specific fields, as e. g. commerce, health, public safety and education. In almost all countries, this legislation is based on the SI. The *International Organisation of Legal Metrology (OIML)* is in charge of the international harmonization of the technical specifications of these legal regulations.

---

**18** The official SI brochure is traditionally available only in French and English. However, translations into various other languages are usually published by the responsible authorities in several countries.

**19** These values are based on the latest values measured in 2018 and published by CODATA on 20th May 2019, but now have *by definition* no longer any measurement uncertainty.

## 3.3 Definition of the base units

The revision of the SI in 2019 has changed the wording of the definitions for all seven base units. However, for the unit of length, the *metre*, the unit of time, the *second*, and the unit of luminous intensity, the *candela*, the physical concept underlying the respective definition has not changed.

The basic concept of the new definitions of base units is that all base units are now defined by relations of values of the *defining constants* listed in table 3.1.

**Tab. 3.1:** The seven *defining constants*.

| Name | Symbol | Value |
|---|---|---|
| hyperfine transition frequency of $^{133}$Cs | $\Delta\nu_{Cs}$ | 9 192 631 770 Hz [a] |
| speed of light in a vacuum | $c$ | 299 792 458 m s$^{-1}$ |
| Planck constant | $h$ | 6.626 070 15 $\times$ 10$^{-34}$ J s [b] |
| elementary charge | $e$ | 1.602 176 634 $\times$ 10$^{-19}$ C [c] |
| Boltzmann constant | $k_B$ | 1.380 649 $\times$ 10$^{-23}$ J K$^{-1}$ [b] |
| Avogadro constant | $N_A$ | 6.022 140 76 $\times$ 10$^{23}$ mol$^{-1}$ |
| luminous efficacy of radiation | $K_{Cd}$ | 683 lm W$^{-1}$ [d] |

[a] $1\,\text{Hz} := 1\,\text{s}^{-1}$
[b] $1\,\text{J} := 1\,\text{kg}\,\text{m}^2\,\text{s}^{-2}$
[c] $1\,\text{C} := 1\,\text{A}\,\text{s}$
[d] $1\,\text{lm} := 1\,\text{cd}\,\text{sr};\ 1\,\text{sr} := 1$

In the following it is shown how to get from the *defining constants* to the SI base units. The definitions of the units of the amount of substance (*mol*, symbol mol) and the luminous intensity (*candela*, symbol cd) can initially be disregarded when deriving the other base units, since their definition results directly from the determination of the numerical values of the Avogadro constant, $N_A$, and the luminous efficacy of radiation, $K_{Cd}$, respectively, as will be shown.[20]

All *defining constants* are quantity values. Therefore, they may be expressed by the equation[21]

$$|Q| = \{Q\}[Q],$$

where $|Q|$ denotes the quantity value of the quantity $Q$, $\{Q\}$ its numerical value and $[Q]$ its associated unit.

---

20 From a mathematical point of view, the units mol and cd are strictly speaking not base units. The unit mol is merely a—albeit very large—multiple of the unit 1, while the unit cd is a fraction of the derived unit W (watt) for power.

21 Note that this equation differs from the equation $Q = \{Q\}[Q]$ often stated in the literature!

We now write the *defining constants* in the form

$$\Delta\nu_{Cs} = \{\Delta\nu_{Cs}\}_{SI}\, s^{-1},$$
$$c = \{c\}_{SI}\, m\, s^{-1},$$
$$h = \{h\}_{SI}\, kg\, m^2\, s^{-1},$$
$$e = \{e\}_{SI}\, A\, s,$$
$$k_B = \{k_B\}_{SI}\, kg\, m^2\, s^{-2}\, K^{-1},$$

where the subscript SI at the numerical values indicates that the respective numerical value defined in the SI has to be used here, which can be taken from table 3.1. We can also represent the *defining constants* by another base in which their numerical values are all equal to one:

$$\Delta\nu_{Cs} = 1\,[\Delta\nu_{Cs}],$$
$$c = 1\,[c],$$
$$h = 1\,[h],$$
$$e = 1\,[e],$$
$$k_B = 1\,[k_B],$$

where $[\Delta\nu_{Cs}]$, $[c]$, $[h]$, $[e]$ and $[k_B]$ denote the respective base units represented by the other base. In this representation, the units of frequency, velocity, action, charge and entropy have been chosen as base units.

The two representations result in the system of equations

$$[\Delta\nu_{Cs}] = \{\Delta\nu_{Cs}\}_{SI}\, s^{-1},$$
$$[c] = \{c\}_{SI}\, m\, s^{-1},$$
$$[h] = \{h\}_{SI}\, kg\, m^2\, s^{-1},$$
$$[e] = \{e\}_{SI}\, A\, s,$$
$$[k_B] = \{k_B\}_{SI}\, kg\, m^2\, s^{-2}\, K^{-1},$$

which can be solved for the SI units with the result

$$1\,\text{s} = C_\text{s}\frac{1}{\Delta \nu_\text{Cs}}\,,$$

$$1\,\text{m} = C_\text{m}\frac{c}{\Delta \nu_\text{Cs}}\,,$$

$$1\,\text{kg} = C_\text{kg}\frac{h\,\Delta \nu_\text{Cs}}{c^2}\,,$$

$$1\,\text{A} = C_\text{A}\,e\,\Delta \nu_\text{Cs}\,,$$

$$1\,\text{K} = C_\text{K}\frac{h\,\Delta \nu_\text{Cs}}{k_\text{B}}\,,$$

where the base units $[\Delta \nu_\text{Cs}]$, $[c]$, $[h]$, $[e]$ and $[k_\text{B}]$ were in turn replaced by the *defining constants* $\Delta \nu_\text{Cs}$, $c$, $h$, $e$, and $k_\text{B}$. The used constants $C_\text{s}$, $C_\text{m}$, $C_\text{kg}$, $C_\text{A}$, and $C_\text{K}$ are compiled in table 3.2. The values given there are *exact rational numbers*, in contrast to the *rounded values* stated in the SI brochure.

**Tab. 3.2:** The numerical values of the constants used.

| Constant | Definition | Exact numerical value |
|---|---|---|
| $C_\text{s}$ | $\{\Delta \nu_\text{Cs}\}_\text{SI}$ | $9\,192\,631\,770$ |
| $C_\text{m}$ | $\dfrac{\{\Delta \nu_\text{Cs}\}_\text{SI}}{\{c\}_\text{SI}}$ | $\dfrac{656\,616\,555}{214\,137\,47}$ |
| $C_\text{kg}$ | $\dfrac{\{c\}^2_\text{SI}}{\{h\}_\text{SI} \times \{\Delta \nu_\text{Cs}\}_\text{SI}}$ | $\dfrac{366\,838\,848\,464\,0072}{248\,616\,420\,290\,3619} \times 10^{40}$ |
| $C_\text{A}$ | $\dfrac{1}{\{e\}_\text{SI} \times \{\Delta \nu_\text{Cs}\}_\text{SI}}$ | $\dfrac{5}{736\,410\,991\,343\,003\,109} \times 10^{26}$ |
| $C_\text{K}$ | $\dfrac{\{k_\text{B}\}_\text{SI}}{\{h\}_\text{SI} \times \{\Delta \nu_\text{Cs}\}_\text{SI}}$ | $\dfrac{276\,1298}{121\,822\,045\,942\,277\,331} \times 10^{11}$ |

It thus turns out that the SI base units for time, the (*second*, symbol s), length, the (*metre*, symbol m), mass, the (*kilogram*, symbol kg), electric current, the (*ampere*, symbol A), and the thermodynamic temperature, the (Kelvin, symbol K) can be obtained simply by a base transformation from the defining constants if these are understood as another basis of the SI.

It remains to show how the units for the amount of substance, the *mol* (symbol mol), and the luminous intensity, the *candela* (symbol cd), are obtained from their respective *defining constants*. For this purpose we write their *defining constants* in the form

$$N_A = \{N_A\}_{SI}\, mol^{-1},$$

$$K_{Cd} = \{K_{Cd}\}_{SI}\, cd\, sr\, W^{-1}.$$

These equations we solve for the base units we are looking for and subsequently insert the numerical values $\{N_A\}_{SI}$ and $\{K_{Cd}\}_{SI}$. This yields

$$1\, mol = \frac{\{N_A\}_{SI}}{N_A} = \frac{6.022\,140\,76 \times 10^{23}}{N_A}$$

and

$$1\, cd = \frac{K_{Cd}}{\{K_{Cd}\}_{SI}}\, W\, sr^{-1} = \frac{K_{Cd}}{683}\, W\, sr^{-1}.$$

These are the relationships stated in the SI brochure, but we have refrained from additionally expressing the power unit watt (see table 3.3) by the SI base units, as it was done in the SI brochure. However, this does not pose any further difficulty.

Having explained how the SI base units result from the *defining constants*, the definitions taken from the SI brochure are easier to comprehend.

**The second**    *The second, symbol s, is the SI unit of time. It is defined by taking the fixed numerical value of the caesium frequency, $\Delta\nu_{Cs}$, the unperturbed ground-state hyperfine transition frequency of the caesium 133 atom, to be 9 192 631 770 when expressed in the unit Hz, which is equal to $s^{-1}$.*

**The metre**    *The metre, symbol m, is the SI unit of length. It is defined by taking the fixed numerical value of the speed of light in vacuum, c, to be 299 792 458 when expressed in the unit $m\, s^{-1}$, where the second is defined in terms of the caesium frequency $\Delta\nu_{Cs}$.*

**The kilogram**    *The kilogram, symbol kg, is the SI unit of mass. It is defined by taking the fixed numerical value of the Planck constant, h, to be 6.626 070 15 $\times$ 10$^{-34}$ when expressed in the unit J s, which is equal to $kg\, m^2\, s^{-1}$, where the metre and the second are defined in terms of c and $\Delta\nu_{Cs}$.*

**The ampere**   The ampere, symbol A, is the SI unit of electric current. It is defined by taking the fixed numerical value of the elementary charge, e, to be $1.602\,176\,634 \times 10^{-19}$ when expressed in the unit C, which is equal to A s, where the second is defined in terms of $\Delta v_{Cs}$.

**The kelvin**   The kelvin, symbol K, is the SI unit of thermodynamic temperature. It is defined by taking the fixed numerical value of the Boltzmann constant, k, to be $1.380\,649 \times 10^{-23}$ when expressed in the unit $J\,K^{-1}$, which is equal to $kg\,m^2\,s^{-2}\,K^{-1}$, where the kilogram, metre and second are defined in terms of h, c and $\Delta v_{Cs}$.

**The mole**   The mole, symbol mol, is the SI unit of amount of substance. One mole contains exactly $6.022\,140\,76 \times 10^{23}$ elementary entities. This number is the fixed numerical value of the Avogadro constant, $N_A$, when expressed in the unit $mol^{-1}$ and is called the Avogadro number.
The amount of substance, symbol n, of a system is a measure of the number of specified elementary entities. An elementary entity may be an atom, a molecule, an ion, an electron, any other particle or specified group of particles.

**The candela**   The candela, symbol cd, is the SI unit of luminous intensity in a given direction. It is defined by taking the fixed numerical value of the luminous efficacy of monochromatic radiation of frequency $540 \times 10^{12}$ Hz, $K_{cd}$, to be 683 when expressed in the unit $lm\,W^{-1}$, which is equal to $cd\,sr\,W^{-1}$, or $cd\,sr\,kg^{-1}\,m^{-2}\,s^3$, where the kilogram, metre and second are defined in terms of h, c and $\Delta v_{Cs}$.

## 3.4 Derived units

Derived units are defined as products of powers of the base units. If the numerical factor of this product is equal to one, the derived units are called coherent. The base units and the coherently derived units of the SI together form the set of coherent SI units. The word *coherent* in this context means that all equations between the numerical values of the quantities have exactly the same form as the equations between the quantities themselves.

**Tab. 3.3:** SI units with special names and symbols.

| Quantity | Unit name, symbol, and definition | | |
|---|---|---|---|
| angular measure [a] | radian | rad | $1$ |
| solid angular measure [a] | steradian | sr | $1$ |
| frequency | hertz | Hz | $s^{-1}$ |
| force | newton | N | $kg\,m\,s^{-2}$ |
| pressure, stress | pascal | Pa | $kg\,m^{-1}\,s^{-2}$ |
| energy, work, amount of heat | joule | J | $kg\,m^2\,s^{-2}$ |
| power, radiant flux | watt | W | $kg\,m^2\,s^{-3}$ |
| electric charge | coulomb | C | $A\,s$ |
| electric potential difference | volt | V | $kg\,m^2\,s^{-3}\,A^{-1}$ |
| capacitance | farad | F | $kg^{-1}\,m^{-2}\,s^4\,A^2$ |
| electric resistance | ohm | $\Omega$ | $kg\,m^2\,s^{-3}\,A^{-2}$ |
| electric conductance | siemens | S | $kg^{-1}\,m^{-2}\,s^3\,A^2$ |
| magnetic flux | weber | Wb | $kg\,m^2\,s^{-2}\,A^{-1}$ |
| magnetic flux density | tesla | T | $kg\,s^{-2}\,A^{-1}$ |
| inductance | henry | H | $kg\,m^2\,s^{-2}\,A^{-2}$ |
| Celsius temperature | degree Celsius | °C | K |
| luminous flux | lumen | lm | $cd\,sr$ |
| illuminance | lux | lx | $cd\,sr\,m^{-2}$ |
| activity referred to a radionuclide | becquerel | Bq | $s^{-1}$ |
| absorbed dose, kerma | gray | Gy | $m^2\,s^{-2}$ |
| dose equivalent | sievert | Sv | $m^2\,s^{-2}$ |
| catalytic activity | katal | kat | $mol\,s^{-1}$ |

[a] Angular measure and solid angular measure are quantities of dimension number, their unit is the number one, symbol $1$.

Some of the coherently derived units in the SI have special names. They are summarized in table 3.3. Together with the seven base units, they form the core of the SI units. All other derived SI units can be represented as a combination of the base units alone or together with some of these special units.

Each physical quantity has only one coherent SI unit, even if this unit can be expressed in different ways, using special names and symbols. Thus, e. g. the units J or N m are used in practice for work, the units W or V A for electric power, the units Wb or V s for magnetic flux, the units H or V s/A for inductance, the units F or A s/V for electric capacitance and the units lx or $cd/m^2$ for illuminance.

The radian and the steradian, the coherent SI units for the plane angle and the solid angle, are supplementary units. These units are special names and signs for the unit one, symbol 1, which indicate that they are special quantities of dimension number. They merely serve to distinguish otherwise identical units of different quantities (such as e. g. in radiometry the units for radiant flux, W, and radiant intensity, W/sr) or to clarify certain relationships (such as e. g. in chemistry the pH value or in information theory the Shannon, Sh [3, 4]), although the respective quantity value could also be represented by a number alone without a unit symbol.

The radian and the steradian had previously the status of supplementary units in the SI until by a resolution of the CGPM in 1995 there status was changed to derived quantities.[22] From a mathematical point of view, this was an unfortunate decision. Angular measures are not derived quantities but pure numbers, i. e. the status of their units is in fact still that of supplementary units and therefore 1 rad = 1 and 1 sr = 1 still apply.

Quantities and units are not unambiguously assigned to each other since several different quantities can have the same SI unit[23] (there are more quantities than units). For example, J/K is not only the SI unit for heat capacity, but also for entropy. The base unit ampere is the SI unit for both the base quantity electric current and the derived quantity magnetomotive force.[24] It is therefore necessary to state the unit only together with its quantity. This applies not only to scientific and technical texts, but also to measuring instruments (i. e. on the display of the instrument both the unit and the measurand shall be indicated).

For some quantities, certain names and symbols for the units are usually preferred to make it easier to distinguish between different quantities with the same SI unit. The use of this option can give an indication of how the corresponding quantity was defined. For example, the quantity torque is defined as the vector product of a location vector and a force vector, and the quantity work is defined as the scalar product of a force vector and a displacement vector. The commonly used unit for

---

22 Resolution 8 of the 20th CGPM, 1995
23 Mathematically speaking, the assignment of quantities and units is a relation that is not a function, i. e. the assignment is not unambiguously invertible. It is therefore not possible to infer from a unit which quantity it was used for.
24 This conflict could be remedied if a base unit for the magnetic flux would be introduced in the SI.

both quantities is therefore the newton metre, N m. Although torque and work have the same SI unit as energy, the energy unit joule, J, is never used, but always the unit newton metre.

The recommended SI unit of frequency is the hertz, Hz, the recommended SI unit of angular velocity and angular frequency is the radian per second, rad/s, and the recommended SI unit of the activity of a radioactive nuclide is the becquerel, Bq. Although it is formally correct to use the SI unit $s^{-1}$ for all three quantities, the use of the special names and signs emphasizes the different character of the quantities concerned. Therefore, these recommendations should be followed.

In order to avoid misunderstandings, the units of the quantities frequency and angular frequency ought to be distinguished. Since *by definition* their quantity values differ by a factor of $2\pi$, failure to take this fact into account could lead to an error of $2\pi$. It is therefore strongly recommended always to use the unit Hz for the quantity frequency and the unit rad/s for the quantities angular frequency and angular velocity and not the unit $s^{-1}$.

In the field of ionizing radiation, the SI unit Becquerel, Bq, is always used and never the unit of frequency, $s^{-1}$. For the absorbed dose and the equivalent dose, the SI units Gray, Gy, and Sievert, Sv, respectively, are used and not the equivalent unit J/kg. The special units Becquerel, Gray and Sievert were introduced specifically because of the hazards for human health that could result from a confusion of the units $(m/s)^2$ and J/kg, respectively, if these units were used incorrectly to identify the different quantities involved.

Special care must be taken when stating temperatures or temperature differences. A temperature difference of 1 K corresponds to that of 1 °C, but for an absolute temperature the difference of 273.15 K between the zero points of the temperature scales must still be taken into account.

## 3.5 Prefix names and prefix symbols

For decimal multiples or fractions of SI units, the prefix names and prefix symbols listed in table 3.4 may be used, e. g. 1 kW (1 kilowatt) instead of 1000 W or 1 μm (1 micrometre) instead of $10^{-6}$ m. However, they do not apply to units that are not SI units. In many countries the use of prefixes is authorized by law.

Prefix names are written in capital letters using the font of the surrounding text. They are inseparably attached to the names of the units to which they are prefixed, e. g. millimetres, hectolitres, nanovolts, etc.

Prefix symbols, like unit symbols, are always written in upright type (Roman font), regardless of the font used in the surrounding text, without a space between the prefix symbol and the unit symbol. All prefixes for factors greater than 1000 are written in upper case, all others in lower case.

**Tab. 3.4:** Prefix names and prefix symbols for SI units.

| Factor | Name | Symbol | Factor | Name | Symbol |
|---|---|---|---|---|---|
| $10^{24}$ | yotta | Y | 0.1 | dezi | d |
| $10^{21}$ | zetta | Z | $10^{-2}$ | centi | c |
| $10^{18}$ | exa | E | $10^{-3}$ | milli | m |
| $10^{15}$ | peta | P | $10^{-6}$ | mikro [a] | µ |
| $10^{12}$ | tera | T | $10^{-9}$ | nano | n |
| $10^{9}$ | giga | G | $10^{-12}$ | piko | p |
| $10^{6}$ | mega | M | $10^{-15}$ | femto | f |
| $10^{3}$ | kilo | k | $10^{-18}$ | atto | a |
| $10^{2}$ | hecto | h | $10^{-21}$ | zepto | z |
| 10 | deka | da | $10^{-24}$ | yokto | y |

[a] Care must be taken to write the Greek letter µ upright (*not italic*).

The group of characters formed by a unit symbol and a preceding prefix symbol represents a new inseparable unit symbol denoting a multiple or fraction of the corresponding unit. This new unit symbol is therefore treated in the same way as the original unit symbol when calculating with quantity values.

*Examples:*
$$(4.7\,\mu m)^2 = (4.7)^2 \times (10^{-6}\,m)^2 = 2.209 \times 10^{-11}\,m^2$$
$$15\,mg/kg = (15 \times 10^{-3}\,g)/(10^3\,g) = 1.5 \times 10^{-5}$$
$$11.3\,g/cm^3 = (11.3/10^{-6}) \times 10^{-3}\,kg/m^3 = 11\,300\,kg/m^3$$
$$1013.25\,hPa = (1013.25 \times 10^2)\,Pa = 101\,325\,Pa$$

Prefix symbols comprising two or more prefix symbols are not permissible. This rule also applies to two or more prefix names. For the unit of mass, e. g. it is not permissible to write µkg (microkilogram) instead of mg (milligram).

Prefix symbols shall not stand alone or precede the number 1 (symbol for the number one, the unit of numbers). Similarly, prefix names may not stand alone or precede the name of the unit one, i. e. the word "one".

The SI is in principle a coherent system of units. However, when prefix names and prefix symbols are used, the resulting units are no longer coherent because each prefix symbol introduces a numerical factor other than one.

## 3.6 Units in the SI that are not SI units

The SI constitutes an internationally agreed reference on the basis of which all other units are defined. The coherent SI units have the great advantage that no conversion of units is required when inserting quantity values into quantity equations.

**Tab. 3.5:** Units in the SI that are not SI units.

| Quantity | Unit name and symbol | | Quantity value in the SI |
|---|---|---|---|
| duration | minute | min | $1\,\text{min} = 60\,\text{s}$ |
| | hour | h | $1\,\text{h} = 60\,\text{min} = 3600\,\text{s}$ |
| | day | d | $1\,\text{d} = 24\,\text{h} = 1440\,\text{min} = 86\,400\,\text{s}$ |
| length | astronomical unit [a] | au | $1\,\text{au} = 149\,597\,870\,700\,\text{m}$ |
| angular measure | degree | ° | $1° = (\pi/180)\,\text{rad}$ |
| | arc minute | ′ | $1' = (1/60)° = (\pi/10\,800)\,\text{rad}$ |
| | arc second | ″ | $1'' = (1/60)' = (\pi/648\,000)\,\text{rad}$ |
| area | hectare | ha | $1\,\text{ha} = (100\,\text{m})^2 = 10^4\,\text{m}^2$ |
| volume | litre [b] | L, l | $1\,\text{L} = 1\,\text{dm}^3 = 10^3\,\text{cm}^3 = 10^{-3}\,\text{m}^3$ |
| mass | ton | t | $1\,\text{t} = 10^3\,\text{kg}$ |
| | dalton [c] | Da | $1\,\text{u} = 1.660\,539\,066\,60(50) \times 10^{-27}\,\text{kg}$ |
| energy | electronvolt [d] | eV | $1\,\text{eV} = 1.602\,176\,634 \times 10^{-19}\,\text{J}$ |

[a] The quantity value was approved by the XXVIII. General Assembly of the International Astronomical Union (IAU) (Resolution B2, 2012).

[b] Preferably, the symbol L should be used to avoid confusion between the letter symbol l and the symbol 1 for the number one.

[c] The dalton (symbol Da) and the atomic mass unit (symbol $m_u$) are alternative names and symbols for the same unit. It is equal to $1/12$ of the mass of a free carbon isotope $^{12}C$ at rest and in the ground state. The stated quantity value was published in 2019 by CODATA.

[d] The electronvolt is the kinetic energy that an electron acquires when passing through a potential difference of one volt in vacuum. The stated quantity value is exact since the revision of the SI in 2019.

There are many more units that are not SI units. These are either of historical interest or are still used in certain fields (e. g. the barrel in the oil business) or in certain countries (e. g. inches, feet and yards). However, the CIPM sees no reason to continue using these units in modern scientific and technical work.

There are areas in which units are traditionally used which are not SI units but which were previously permitted together with the SI because of their use based on international agreements. These include e. g. the nautical mile (which is equal to $1852\,\mathrm{m}$) used in navigation (for nautical charts according to the *International Hydrographic Organization*, IHO) and in aviation (in accordance with the rules of the *International Civil Aviation Organization*, ICAO, and the *International Air Transport Association*, IATA) to measure distances. The *United Nations Convention on the Law of the Sea (UNCLOS)* is also based on the nautical mile. It is therefore incomprehensible why the nautical mile is no longer permitted within the SI.

## 3.7 Units of logarithmic ratio quantities

Whenever large number ranges are involved, it is easier to calculate by using the logarithms of the numbers instead of the numbers themselves. Therefore, logarithmic ratio quantities were introduced to facilitate scientific and technical work. These quantities are standardized in IEC 80000-15 [5].

**Figure and level**

In electrical engineering and acoustics, the logarithmic ratios of two power quantities or root-power quantities (e. g. electrical voltage, sound pressure) are used.

**Tab. 3.6:** Examples of figures.

| Formula | Denomination | Supplementary unit |
|---|---|---|
| $R = 10\lg\dfrac{I_\mathrm{in}}{I_\mathrm{out}}$ | sound reduction index | dB (decibel) |
| $a = \ln\left\|\dfrac{U_1}{U_2}\right\|$ | voltage attenuation figure | Np (neper) |

Depending on whether the reference quantity used in the calculation of the ratio is a variable quantity or a fixed quantity, a distinction is made between the two logarithmic ratio quantities *figure* (e. g. gain figure, noise figure) and *level* (e. g. sound pressure level).

For logarithmic quantities, but only for these, the supplementary unit[25] Bel (unit symbol B) or its tenth part, the decibel (unit symbol dB), is employed when using the decadic logarithm, and the supplementary unit neper (unit symbol Np) when using the natural logarithm.

A level is a logarithmic quantity defined by the logarithm of the ratio of a power quantity or a root-power quantity to a fixed reference quantity of the same kind.

**Tab. 3.7:** Examples of levels.

| Formula | Denomination | Reference | Supplementary unit |
|---------|--------------|-----------|--------------------|
| $L = \ln \dfrac{P}{P_0}$ | power level | $P_0$ | Np (neper) |
| $L = 10 \lg \dfrac{P}{P_0}$ | power level | $P_0$ | dB (decibel) |
| $L_U = 20 \lg \dfrac{U}{U_0}$ | voltage level | $U_0$ | dB (decibel) |

If the difference of two levels with the same reference quantity[26] is calculated, the result does not depend on the reference quantity, because e. g.

$$\Delta L = L_2 - L_1 = \ln \frac{P_2}{P_0} - \ln \frac{P_1}{P_0} = \ln \frac{P_2}{P_1}$$

applies to the difference of power levels. The quantity $\Delta L$ is not a level but a figure, since the quantity in the denominator of the logarithmic ratio is no longer a fixed reference quantity.

**pH value**

The pH value is a measure of the acidic or basic character of an aqueous solution. It is defined as the negative decadic logarithm of the hydrogen ion activity:

$$pH = - \lg a(H^+).$$

The activity $a$ is a thermodynamic quantity of dimension number, which is used in physical chemistry instead of the concentration of substances. The pH value is a quantity of dimension number, i. e. its unit is the number one.

---

25 Supplementary units are special names and signs for the unit one or the symbol 1, indicating that they are special quantities of dimension number

26 This difference used to be called "relative level". Since this name is misleading (it is not a quotient, but a difference), it should no longer be used

## Quantities of information

Quantities of information are used to quantify information. The information content (also called self-information or surprisal) is defined as the negative logarithm of the probability of an event, i. e.

$$I(E) = -\log_a p(E)$$

is valid, where $I(E)$ denotes the information content of the event $E$ and $p(E)$ its probability. The base of the logarithm is denoted by $a$.

The information content is a quantity of dimension number, but three different supplementary units are used in order to indicate the base of the logarithm used. For the commonly used base 2, the standardized supplementary unit is the shannon (unit symbol Sh).[27]

If the decadic logarithm is used for the information content, the supplementary unit is called hartley (unit symbol hart) and if the natural logarithm is used, it is called nat (unit symbol nat).

The so-called information entropy introduced in 1948 by Claude Shannon[28]

$$H = -\sum_{i=1}^{n} p_i \operatorname{lb} p_i$$

is also a quantity of information. It originally served only to measure the information content of a message consisting of $n$ symbols occurring with the probabilities $p_i$ $(p_i = 1, \ldots, n)$. Today, this information measure is also used to measure the entropy of stochastically occurring events.

In information theory also other information measures exist (such as e. g. the differential entropy), but these will not be discussed here. For more information, consult the special literature on information theory.

## Storage capacity

Closely related to the quantities of information is the storage capacity

$$M_e = \operatorname{lb} n,$$

where $n$ denotes the number of states of a memory.

---

27 The shannon was introduced by the ISO in 1975, but in practice it cannot prevail against the previously valid supplementary unit bit (unit symbol bit).

28 The information entropy is mathematically related to the definition of the entropy introduced in 1875 in statistical thermodynamics, from which it differs only by the Boltzmann constant $k_B$.

The storage capacity is a quantity of dimension number for which the supplementary unit bit (unit symbol bit, occasionally also b) is usually used.

In connection with the storage capacity, not only the prefix names and prefix symbols of the SI are used, but also the binary prefix names and prefix symbols according to IEC 80000-13 [3] listed in table 3.8, which were additionally introduced by the IEC in 1996. The difference between the prefixes as standardized in ISO and IEC is often unknown or incorrectly applied or interpreted. In particular, this often leads to confusion among users of mass storage media.

**Tab. 3.8:** Prefix names and prefix symbols according to IEC 80000-13.

| Factor | Name | Symbol | Value |
|---|---|---|---|
| $2^{80}$ | yobi | Yi | 1 208 925 819 614 629 174 706 176 |
| $2^{70}$ | zebi | Zi | 1 180 591 620 717 411 303 424 |
| $2^{60}$ | exbi | Ei | 1 152 921 504 606 846 976 |
| $2^{50}$ | pebi | Pi | 1 125 899 906 842 624 |
| $2^{40}$ | tebi | Ti | 1 099 511 627 776 |
| $2^{30}$ | gibi | Gi | 1 073 741 824 |
| $2^{20}$ | mebi | Mi | 1 048 576 |
| $2^{10}$ | kibi | Ki | 1024 |

In computer science, the supplementary unit byte (unit symbol B) is also used for the storage capacity, where $1\,B = 8$ bit. Historically, the byte was used to encode a single character of text in a computer. Therefore, it is still the smallest addressable unit of memory in most computers today. Modern computer architectures use 32-bit or 64-bit words, i. e. four or eight bytes, respectively.

The supplementary unit byte is also used together with prefix names and prefix symbols. Both the SI convention (byte is *not* a supplementary unit in the SI) and the IEC convention are used. Unfortunately, this frequently leads to misunderstandings because the correct prefixes are not used.

The manufacturers of mass storage media, such as hard disks, memory chips, blank DVDs and USB memory sticks usually apply the SI convention. Operating systems use one of the two conventions or give the user a choice. The Microsoft operating systems use the IEC convention internally, but unfortunately do not use the correct prefix symbols when displaying the storage capacity. Thus e. g. for a hard disk with a storage capacity of 1 TB only 931 GB is displayed, but actually the approximate value 931 GiB is meant. For blank CDs, the storage capacity is incorrectly indicated as 700 MB instead of 700 MiB, whereas for blank DVDs it is correctly indicated as 4.7 GB, which makes confusion complete.

**Absorbance**

The absorbance is a measure of the attenuation of electromagnetic radiation (e. g. light) when passing through a medium. It is defined as the negative decadic logarithm of the spectral transmittance $\tau(\lambda)$:

$$E_\lambda = -\lg \tau(\lambda).$$

The absorbance depends on the wavelength of the radiation. It is a quantity of dimension number, i. e. its unit is the number one.

**Frequency level**

The frequency level is a logarithmic quantity used in acoustics, which is defined in ISO 80000-8 [6] as the binary logarithm of a frequency ratio

$$m = \mathrm{lb}\, \frac{f_1}{f_2},$$

where $m$ denotes the frequency interval and $f_1$ and $f_2$ denote frequencies. The frequency interval is a quantity of dimension number for which the supplementary unit octave (unit symbol oct) is used.

Logarithms have been used in music since the time of HUYGENS, because they facilitate the comparison between calculation and listening experience due to the approximate logarithmic auditory sensation. LEONHARD EULER suggested to use the base 2 for the logarithm of frequency ratios, whereby equal intervals of different octaves have equal mantissas and differ only in the index number; the octave is assigned the value 1.

The frequency level is also used in high-frequency engineering to indicate frequency ranges and is particularly common in antenna technology. Here, however, instead of the binary logarithm, the decadic logarithm with the corresponding supplementary unit decade (unit symbol dec) is preferred.

## 3.8 Calculating with units

Units can be multiplied and divided. These arithmetic operations also include exponentiation with integer exponents. Fractional exponents are not allowed, however, because the result does not make sense.

For the notation of the multiplication of two units, one of the following options can be used:

$$\mathrm{N\,m,} \qquad \mathrm{N \cdot m.}$$

For the notation of the division of a unit by another unit, one of the following options can be used:

$$\frac{\mathrm{m}}{\mathrm{s}}, \qquad \mathrm{m/s}, \qquad \mathrm{m\,s^{-1}}, \qquad \mathrm{m \cdot s^{-1}}.$$

Not more than one division symbol shall be used in an expression.

*Examples*:   not: cm/s/s    but: $\mathrm{cm/s^2}$    or   $\mathrm{cm\,s^{-2}}$
not: J/K/mol   but: J/(K mol)   or   $\mathrm{J\,K^{-1}\,mol^{-1}}$

Since the rules of algebra can be applied not only to pure numbers but also to units and quantity values, it is permissible to divide a quantity value by its unit. Therefore the expression

$$\{Q\} = \frac{|Q|}{[Q]}$$

is valid, where $|Q|$ denotes the quantity value of a quantity $Q$, $[Q]$ the unit and $\{Q\}$ the numerical value of the quantity. Numerical values are listed in tables or used when labelling axes of coordinate systems. The expression of "quantity value/unit" shall be used in the headings of tables and as labels at the axes in figures to indicate uniquely the meaning of the numbers to which it refers. An example for the application of these rules is shown in the figure.

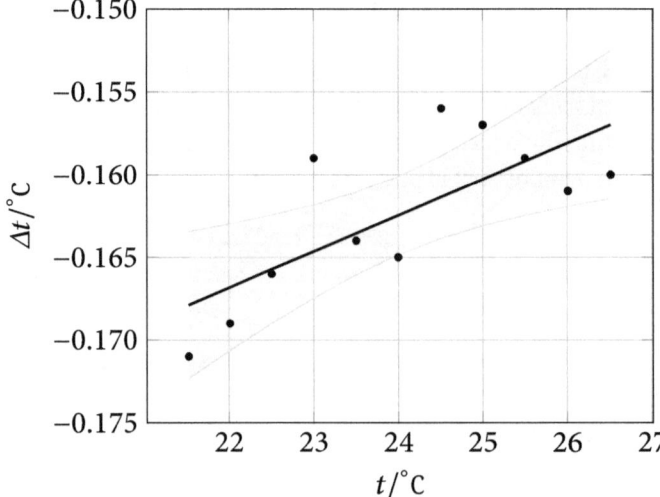

| $t/°C$ | $\Delta t/°C$ |
|--------|---------------|
| 21.521 | -0.171 |
| 22.012 | -0.169 |
| 22.512 | -0.166 |
| 23.003 | -0.159 |
| 23.507 | -0.164 |
| 23.999 | -0.165 |
| 24.513 | -0.156 |
| 25.002 | -0.157 |
| 25.503 | -0.159 |
| 26.010 | -0.161 |
| 26.511 | -0.160 |

## 3.9 Exact physical constants

Only three of the *defining constants* of the SI are fundamental constants of nature. Constants of nature are spatially and temporally invariant quantities. They are part of physical theories, but their values cannot be determined by the theories themselves, but must in principle be determined by measurements with respect to arbitrarily chosen units, preferably defined within a system of units.

The base units of a system of units can, however, also be determined in such a way that the values of certain constants of nature have fixed values in the chosen system of units. This possibility was used in the revision of the SI in 2019 by fixing the values of the fundamental constants of nature, the speed of light in vacuum $c$, the Planck constant $h$ and the elementary charge $e$, with respect to the already existing SI base units. Hence, all physical constants derived from these fundamental constants of nature have now also an exact value in the SI.

The hyperfine transition frequency $\Delta\nu_{Cs}$ of $^{133}$Cs can also be regarded as a derived physical constant. Its numerical value was determined by measurements in the years 1955 to 1958 [29] and is the basis of the still valid definition of the unit of time in the SI adopted in 1967 by the *General Conference on Weights and Measures (CGPM)*. The numerical value of the frequency $\Delta\nu_{Cs}$ has since been fixed by agreement as exact. However, with a new definition of the unit of time to be expected in the near future, a different frequency will be the basis for the unit of time and will then be agreed to be exact as well. Then the hyperfine transition frequency of the $^{133}$Cs must again be determined by measurement with respect to this new frequency.

The Avogadro constant, $N_A$, has today an exact numerical value determined by agreement in such a way that the continuity of the measurement results with respect to the unit of the amount of substance before and after the redefinition of the SI in 2019 was preserved. Until the redefinition, a mole was defined as the amount of substance consisting of as many particles as there are atoms in 12 g of the isotope $^{12}$C in the ground state.[30] The atomic mass unit $m_u$ and the mole were thus defined by the same isotope, i. e. $1\,m_u N_A = 1$ g/mol was valid, and the numerical value of the Avogadro constant was determined by measurement.

Today, the numerical value of the Avogadro constant is exactly fixed by agreement and the atomic mass unit must be determined by measurement. The currently valid value is $1\,m_u = 1.660\,539\,066\,60(50) \times 10^{-27}$.

---

29 The relative uncertainty was reported to be about $2 \times 10^{-9}$
30 This corresponds to the decision of the *General Conference on Weights and Measures (CGPM)* in 1971 when introducing the SI base unit mole for the amount of substance.

By a resolution of the General Conference on Weights and Measures (CGPM) in 1948, it was decided that the absolute thermodynamic scale should have the triple point of water as the only fundamental fixed point, and in 1954 the thermodynamic temperature of the triple point of water was then fixed by a resolution of the CGPM at exactly 273.16 K and thus the unit of thermodynamic temperature [31] was defined. The Boltzmann constant $k_B$ became a constant of proportionality between the thermal energy $E_{th} = k_B T$ and the derived energy unit joule already defined in the SI. Its numerical value had to be determined by measurement. Today, the numerical value of the Boltzmann constant is exactly fixed by agreement and the thermodynamic temperature of the triple point of water must be determined by measurement.

The most important exact derived physical constants in the SI that are valid since the redefinition of the SI due to the fixed numerical values of the fundamental constants of nature, the speed of light in vacuum $c$, the Planck constant $h$ and the elementary charge $e$, as well as the Avogadro constant $N_A$ and the Boltzmann constant $k_B$, are summarized in the following table.

**Tab. 3.9:** Exact derived physical constants in the SI.

| Quantity | Symbol | Definition |
|---|---|---|
| magnetic flux quantum | $\Phi_0$ | $h/(2e)$ |
| Josephson constant | $K_J$ | $2e/h$ |
| von Klitzing constant | $R_K$ | $h/e^2$ |
| conductance quantum | $G_0$ | $2e^2/h$ |
| first radiation constant | $c_1$ | $2\pi h c^2$ |
| molar gas constant | $R$ | $k_B N_A$ |
| Faraday constant | F | $e N_A$ |
| Stefan-Boltzmann constant | $\sigma$ | $(2\pi^5/15)k_B^4/(h^3 c^2)$ |
| second radiation constant | $c_2$ | $hc/k_B$ |

It should also be noted that due to the redefinition of the ampere, the value of the magnetic constant $\mu_0$ is no longer exactly equal to $4\pi \times 10^{-7}$ N/A$^2$, but must be measured. Due to the relationship $\mu_0 \varepsilon_0 c^2 = 1$, this also applies to the electric constant $\varepsilon_0$, which was exact before the redefinition of the ampere, but today must be measured as well.

---

**31** The unit was initially the "degree kelvin", unit symbol °K, and was changed in 1967 to kelvin, unit symbol K.

# 4 Notation of numbers

## 1. Number, number symbol, number value

A *number* is denoted by a *number symbol* representing a *number value*. The number symbol consists of one or more *digits* written *upright* and, if necessary, additional characters, such as e. g. a sign and a decimal separator.

## 2. Sign

A minus sign (−) may be placed in front of a number symbol to indicate a negative numerical value.

## 3. Decimal separator

According to ISO the *decimal separator* is generally a comma on the line (,). In English documents, a dot on the line (.) may also be used. The half-height dot (·), historically used in British texts, should never be used as a decimal separator.

There shall always be at least one digit before and after the decimal separator. An integer shall never be terminated by a decimal separator. If the absolute value of a number is less than one, a decimal separator shall be preceded by a zero.

| *Examples*: | 35 or 35.0 | *but not* | 35. |
|---|---|---|---|
| | 0.0035 | *but not* | .0035 |

## 4. Digit grouping

To facilitate the reading of long numbers (more than four digits either to the right or to the left of the decimal separator), the digits may be grouped into groups of three separated by a thin space; however, except the decimal separator, no comma or full stop shall be used. Instead of a single final digit, the last four digits may also be grouped together.

https://doi.org/10.1515/9783111344119-004

*Examples*:   1987
299 792.458
1.234 567 8
1.234 5678

## 5. Multiplication

The *multiplication sign* between numbers is a half-height dot (·).[32] If a dot on the line is used as a decimal separator in English texts, a cross (×) shall be used as multiplication sign or the numbers shall be bracketed.

*Examples*:   2.3 · 3.4
2.3 × 3.4   or   (2.3) × (3.4)

## 6. Division

The *division* of two numbers can be represented either by a horizontal line (this is preferable in formulas) or by the division symbol (/).

*Examples*:   $$\frac{136}{273.15}$$   or   136/273.15

In certain cases, it can also be represented as the product of the numerator and the negative first power of the denominator. In this case, the number with the negative exponent shall always be placed in parentheses:

*Example*:   $136 \times (273.15)^{-1}$

If the division symbol (/) is used and it is doubtful where the numerator begins or the denominator ends, a bracket shall be used.

*Example*:   13/27 × 4   *means*   13/(27 × 4)   *or*   (13/27) × 4 ???

---

**32** This is the ISO recommendation, but in English texts a cross (×) can also be used.

# 5 Mathematical symbols

A comprehensive list of mathematical signs and symbols with definitions and explanations can be found in ISO 80000-2 [8].

## 5.1 General symbols

**Tab. 5.1:** General symbols

| | |
|---|---|
| $=$ | equal to |
| $\neq$ | not equal to |
| $\equiv$ | identically equal to, equivalent to |
| $:=\,,\ =:\,,\ \stackrel{\text{def}}{=}\,,\ =_{\text{def}}$ | by definition equal to |
| $\hat{=}$ | corresponds to |
| $\approx$ | approximately equal to |
| $\simeq$ | asymptotically equal to |
| $\sim,\ \propto$ | proportional to |
| $\longrightarrow$ | approaches |
| $>$ | greater than |
| $<$ | less than |
| $\gg$ | much greater than |
| $\ll$ | much less than |
| $\geqslant,\ \geqq,\ \geq$ | greater than or equal to, at least equal to |
| $\leqslant,\ \leqq,\ \leq$ | less than or equal to, at most equal to |
| $+$ | plus |
| $-$ | minus |
| $\pm$ | plus or minus |
| $a\,b,\ a\cdot b,\ a\times b$ | $a$ multiplied by $b$ |
| $a/b,\ \dfrac{a}{b},\ a\,b^{-1}$ | $a$ divided by $b$ |
| $a^{n}$ | $a$ raised to the power $n$ |
| $|a|$ | absolute value of $a$ |

https://doi.org/10.1515/9783111344119-005

**Tab. 5.1:** continued

| | |
|---|---|
| $\lfloor a \rfloor$ | greatest integer less than or equal to $a$ |
| $\lceil a \rceil$ | least integer greater than or equal to $a$ |
| $\sqrt{a},\ a^{1/2}$ | square root [†] of $a$ |
| $\sqrt[n]{a},\ a^{1/n}$ | $n$-th root of $a$ |
| $\bar{a},\ \langle a \rangle$ | mean (value) of $a$ |
| $n!$ | $n$ factorial |
| $\binom{n}{k}$ | binomial coefficient ($n$ over $k$); $\quad \binom{n}{k} = \dfrac{n!}{k!(n-k)!}$ |
| $\infty$ | infinity |

[†] The notation $\sqrt{}(x + y)$, often used in older English literature, shall not be used instead of $\sqrt{x + y}$.

## 5.2 Letter symbols

Letter symbols and symbolic letter representations of mathematical operations should be written *upright* (not in *italic* type).

**Tab. 5.2:** Letter symbols

| | |
|---|---|
| $\pi$ | ratio of the circumference of a circle to its diameter |
| e | base of the natural logarithm |
| $\exp(x)$, $e^x$ | exponential of $x$ |
| $\log_a x$ | logarithm of $x$ to the base $a$ |
| $\ln x$, $\log_e x$ | natural logarithm of $x$ |
| $\lg x$, $\log_{10} x$ | common (decadic) logarithm of $x$ |
| $\operatorname{lb} x$, $\log_2 x$ | binary logarithm of $x$ |
| $\sum$ | sum |
| $\prod$ | product |
| $\Delta x$ | finite increment $^\dagger$ of $x$ |
| $\delta x$ | variation of $x$ |
| $\mathrm{d}x$ | total differential of $x$ |
| $f(x)$ | function (of the variable) $x$ |
| $g \circ f$ | composite function of $f$ and $g$;   $g \circ f = g(f(x))$ |
| $\lim_{x \to a} f(x)$, $\lim_{x \to a} f(x)$ | limit of $f(x)$ if $x$ approaches $a$ |
| $\dfrac{\mathrm{d}f}{\mathrm{d}x}$, $\mathrm{d}f/\mathrm{d}x$, $f'$ | derivative of $f$ |
| $\dfrac{\partial f}{\partial x}$, $\partial f/\partial x$, $\partial_x f$, $f_x$ | partial derivative of $f$ |
| $\mathrm{d}f$ | total differential of $f$;   $\mathrm{d}f = \left(\dfrac{\partial f}{\partial x}\right)\mathrm{d}x + \left(\dfrac{\partial f}{\partial y}\right)\mathrm{d}y$ |
| $\delta f$ | variation of $f$ |
| $\delta(x)$, $\delta(r)$ | Dirac delta function;   $\delta(r) = \delta(x)\delta(y)\delta(z)$ |
| $\delta_{ik}$ | Kronecker delta symbol;   $\delta_{ik} = \begin{cases} 1 & \text{if } i = k \\ 0 & \text{if } i \neq k \end{cases}$ |
| $\operatorname{sgn} a$ | signum of $a$;   $\operatorname{sgn} a = \dfrac{a}{\|a\|}$, $a \neq 0$ |

$^\dagger$ upright Greek delta, not a triangle

## 5.3 Trigonometric and hyperbolic functions

Letter representations of trigonometric and hyperbolic functions shall be written *upright* (not in *italic* type).

**Tab. 5.3:** Trigonometric and hyperbolic functions

| | |
|---|---|
| $\sin x$ | sine of $x$ |
| $\cos x$ | cosine of $x$ |
| $\tan x$ | tangent [a] of $x$ |
| $\cot x$ | cotangent [b] of $x$ |
| $\sec x$ | secant of $x$; $\sec x = \dfrac{1}{\cos x}$ |
| $\csc x$ | cosecant of $x$; $\csc x = \dfrac{1}{\cos x}$ |

[a] The notation tg $x$ is deprecated and shall not be used instead of tan $x$.
[b] The notation ctg $x$ is deprecated and shall not be used instead of cot $x$.

The trigonometric functions $\sec x$ and $\csc x$ are less often used.

Note that the functions $\sec x$ and $\csc x$ are *not* identical with the inverse trigonometric functions $\cos^{-1}x$ and $\sin^{-1}x$, respectively.

For the *inverse functions of the trigonometric functions*, the symbolic representation of the corresponding trigonometric function prepended by the three letters "arc" is recommended.

    *Examples*:    $\arcsin x$, $\arccos x$, $\arctan x$,   etc.

For the *hyperbolic functions* the symbolic representation of the corresponding trigonometric function with the letter "h" appended is recommended.

    *Examples*:    $\sinh x$, $\cosh x$, $\tanh x$,   etc.

For the *inverse functions of hyperbolic functions* the symbolic representation of the corresponding hyperbolic functions prepended by the two letters "ar" is recommended.

    *Examples*:    arsinh $x$, arcosh $x$, artanh $x$,   etc.

## 5.4 Periodically time-dependent quantities

The temporal mean value of $x(t)$ is defined by

$$\overline{x} = \frac{1}{T} \int\limits_0^T x(t)\,\mathrm{d}t$$

and the temporal root-mean-square value of $x(t)$ is defined by

$$\tilde{x} = \left( \frac{1}{T} \int\limits_0^T |x(t)|^2 \mathrm{d}t \right)^{1/2}.$$

The denotations of the values of periodically time-dependent quantities are shown in the following table.

**Tab. 5.4:** Denotations of periodically time-dependent quantities.

| Value type | case A | case B |
|---|:---:|:---:|
| instantaneous value | $x$ | $x$ |
| mean value | $\overline{x}, \langle x \rangle$ | $\overline{x}, \langle x \rangle$ |
| root-mean-square value [a] | $\tilde{x}$ | $X$ |
| maximum value, peak value [b] | $\hat{x}$ | $\hat{x}, \hat{X}$ |
| minimum value [c] | $\check{x}$ | $\check{x}, \check{X}$ |

[a] The symbols $x_{rms}$ and $x_{eff}$ are also used.
[b] The maximum value of $x$ can also be denoted by $x_{max}$.
[c] The minimum value of $x$ can also be denoted by $x_{min}$.
*Note*: Case A applies if only a lower case letter or only an upper case letter shall be used to denote the quantity.
Case B applies if both a lower-case letter and an upper-case letter can be used to denote the quantity.

## 5.5 Vector- and tensor calculus

**Tab. 5.5:** Symbols for vector- and tensor calculus

| | |
|---|---|
| $a, A, \vec{a}, \vec{A}$ | vector |
| $|a|$ | absolute value of $a$ |
| $e_a$ | unit vector;   $e_a = a/|a|$ |
| $e_x, e_y, e_z; i, j, k$ | unit coordinate vectors [a] |
| $a \cdot b$ | scalar product (inner product) of $a$ and $b$ |
| $a \times b$ | vector product [b] of $a$ and $b$ |
| $a \wedge b$ | outer product (Grassmann product) of $a$ and $b$ |
| $ab$ | dyadic product of $a$ and $b$ |
| $a \otimes b$ | tensor product of $a$ and $b$ |
| $\nabla, \vec{\nabla}$ | differential vector operator, nabla [c] |
| $\mathrm{grad}\,\varphi, \nabla\varphi$ | gradient of a scalar field $\varphi$ |
| $\mathrm{div}\,A, \nabla \cdot A$ | divergence of a vector field $A$ |
| $\mathrm{curl}\,A, \mathrm{rot}\,A, \nabla \times A$ | curl of a vector field $A$ |
| $\Delta\varphi, \nabla^2\varphi$ | Laplacian [d] |
| $\Box\,\varphi$ | d'Alembertian |
| $A$ | second order tensor |
| $S:T$ | scalar product of $S$ and $T$;   $S:T = \sum_{i,k} S_{ik} T_{ki}$ |
| $S \cdot T$ | tensor product of $S$ and $T$;   $(S \cdot T)_{ij} = \sum_{k} S_{ik} T_{kj}$ |
| $S \cdot a$ | product of tensor $S$ and vector $a$;   $(S \cdot a)_i = \sum_{k} S_{ik} a_k$ |

[a] The symbols $1_x, 1_y, 1_z$ are also used.
[b] The notation $a \wedge b$ shall not be used instead of $a \times b$.
[c] The notation $\partial/\partial r$ shall not be used instead of the nabla operator.
[d] The notation $\nabla\nabla$ can also be used instead of $\nabla^2$.

## 5.6 Matrix calculus

**Tab. 5.6:** Symbols for matrix calculus

| | |
|---|---|
| $A, (a)_{ik}$ | matrix; $\quad A = \begin{pmatrix} a_{11} & \cdots & a_{1n} \\ \vdots & \ddots & \vdots \\ a_{m1} & \cdots & a_{mn} \end{pmatrix}$ |
| $AB$ | product of $A$ and $B$ |
| $A^{-1}$ | inverse of $A$ |
| $I, E$ | unit matrix |
| $A^{\mathsf{T}}, \tilde{A}$ | transpose of $A$; $\quad A^{\mathsf{T}}_{ik} = A_{ki}$ |
| $A^*$ | complex conjugate of $A$; $\quad A^*_{ik} = (A_{ik})^*$ |
| $A^{\dagger}$ | Hermitian conjugate of $A$; $\quad A^{\dagger}_{ik} = (A_{ki})^*$ |
| $\det A$ | determinant of $A$ |
| $\operatorname{tr} A$ | trace of $A$ |

## 5.7 Complex quantities

**Tab. 5.7:** Symbols for complex quantities

| | |
|---|---|
| $i, j$ [†] | imaginary unit; $\quad i^2 = -1$ |
| $\mathbb{R}z, z'$ | real part of $z$ |
| $\mathbb{J}z, z''$ | imaginary part of $z$ |
| $|z|$ | modulus of $z$ |
| $\arg z, \varphi$ | argument of $z$, phase of $z$; $\quad z = |z|\, e^{i\varphi}$ |
| $z^*, \bar{z}$ | complex conjugate of $z$ |

[†] The symbol $j$ is often used in electrical engineering formulas to avoid confusion with the symbol $i$, which is commonly used there to denote the quantity alternating electric current.

## 5.8 Symbols for set theory

**Tab. 5.8:** Set theory

| | |
|---|---|
| $\in$ | is an element of; $x \in M$ |
| $\notin$ | is not an element of; $x \notin M$ |
| $\ni$ | contains as element; $M \ni x$ |
| $\{a_1, a_2, \ldots, a_n\}$ | set of elements $a_1 \ldots a_n$ |
| $\{x \in M \mid p(x)\}$ | set of elements of $M$ for which $p(x)$ is true |
| $\emptyset, \{\}$ | empty set |
| $\mathbb{N}$ | the set of natural numbers [†] |
| $\mathbb{N}_0$ | the set of natural numbers including the number 0; $$\mathbb{N}_0 := \mathbb{N} \cup \{0\}$$ |
| $\mathbb{Z}$ | the set of integer numbers |
| $\mathbb{Q}$ | the set of rational numbers |
| $\mathbb{R}$ | the set of real numbers |
| $\mathbb{C}$ | the set of complex numbers |
| $\subseteq, (\subset)$ | subset; $A \subseteq B$ |
| $\subset, (\subsetneq)$ | proper subset; $A \subset B$ |
| $\supseteq, (\supset)$ | contains; $A \supseteq B$ |
| $\supset, (\supsetneq)$ | contains properly; $A \supset B$ |
| $\cup$ | union; $A \cup B = \{x \mid x \in A \vee x \in B\}$ |
| $\cap$ | intersection; $A \cap B = \{x \mid x \in A \wedge x \in B\}$ |
| $\setminus$ | difference; $A \setminus B = \{x \mid x \in A \wedge x \notin B\}$ |
| $A^C, \overline{A}$ | complement of $A$; $A^C = \{x \mid x \notin A\}$ |
| $A \times B$ | Cartesian product of $A$ and $B$ |
| $\mathcal{P}(M), \mathfrak{P}(M)$ | power set of $M$ |

[†] $\mathbb{N}$ may contain the number 0.

## 5.9 Mathematical logic

**Tab. 5.9:** Symbols for mathematical logic

| | |
|---|---|
| ⊤ | tautology |
| ⊥ | contradiction |
| ∧ | conjunction (and) |
| ∨ | disjunction (inclusive or) |
| ⊻, (≢, ⊕) | exclusive disjunction (exclusive or) |
| ¬ | negation (not) |
| ⇒, → | material implication (if ... , then ... ) |
| ⇔, ↔ | equivalence, ( ... if and only if ... ) |
| ∀ | universal quantifier (for all ... ) |
| ∃ | existential quantifier (there is at least one ... ) |
| ∃! | uniqueness (there is exactly one ... ) |
| ⊢ | proves (syntactically entails) |
| ⊨ | models (semantically entails) |

# 6 Symbols of physical quantities

This chapter contains tables of the commonly used symbols of physical quantities. They are intended to serve as a guide for teachers and students and to facilitate the exchange of information across disciplinary boundaries. However, it was not intended to include all known symbols used in physics, chemistry and engineering. For a considerably more comprehensive compilation of quantity symbols with corresponding descriptions, see the ISO 80000 series of standards. The number of quantities used in scientific and technical fields is in principle unlimited. Therefore, it is impossible to provide a complete list of quantities.

Each symbol is listed in the category that seems most appropriate and is generally repeated in another category only if such repetition seems useful for a logical grouping of related symbols. The emphasis here is on symbols and nomenclature, e. g. a formula given together with the name of a symbol shall not be regarded as a definition but only as a description.

Many of the symbols listed here are generic. They may be further specified, if that should prove necessary, by using upper and lower case letters or by adding superscript or subscript letters, provided that this does not create ambiguities or conflicts with other symbols.

If more than one symbol is given in the tables, there is no implicit preference for their order. Symbols in brackets are usually alternative options available to avoid the repeated use of a symbol with different meanings. In cases where there are several forms for a Greek letter ($\varepsilon$, $\epsilon$; $\vartheta$, $\theta$; $\kappa$, $\varkappa$; $\phi$, $\varphi$), any of them may be used. The form $\varpi$ of the letter $\pi$ can also be used if appropriate.

There are two different conventions for writing vectors in physics. In the tables, vectors are written in bold type without an arrow. However, this does not exclude the representation with an arrow.

https://doi.org/10.1515/9783111344119-006

# 6.1 Space and time

**Tab. 6.1:** Quantities of space and time

| | |
|---|---|
| $(x, y, z), (x_1, x_2, x_3)$ | Cartesian coordinates |
| $(r, \varphi)$ | polar coordinates |
| $(r, \varphi, z)$ | cylinder coordinates |
| $(r, \vartheta, \varphi)$ | spherical coordinates |
| $r$ | position vector |
| $l, L, a$ | length |
| $b$ | width |
| $h$ | height |
| $r$ | radius |
| $d, \delta$ | thickness |
| $d$ | diameter; $\quad d = 2r$ |
| $ds, dl$ | path element |
| $A, S$ | area, area of a surface |
| $V, (v)$ | volume |
| $\alpha, \beta, \gamma, \theta, \vartheta, \phi$ | angular measure (of a plane angle, angle of rotation or phase angle) |
| $\omega, \Omega$ | solid angular measure |
| $\lambda$ | wavelength |
| $\sigma, (\tilde{\nu})$ | wave number; [†] $\sigma = 1/\lambda$ |
| $\sigma$ | wave vector |
| $k$ | angular wave number; $\quad k = 2\pi/\lambda$ |
| $k$ | propagation vector, angular wave vector |
| $\alpha$ | damping coefficient |
| $\beta$ | phase coefficient |
| $\gamma$ | propagation constant; $\quad \gamma = \alpha + i\beta$ |
| $t$ | time, duration, period |
| $T$ | period duration, period |
| $\nu, f$ | frequency; $\quad \nu = 1/T$ |
| $\omega$ | circular frequency, angular frequency; $\quad \omega = 2\pi\nu$ |

**Tab. 6.1:** continued

| | |
|---|---|
| $\tau$ | relaxation (decay) time, time constant; $\quad F(t) = e^{-t/\tau}$ |
| $\delta$ | decay coefficient; $\quad F(t) = e^{-\delta t} \sin \omega t$ |
| $\gamma$ | growth rate; $\quad F(t) = e^{\gamma t} \sin \omega t$ |
| $\Lambda$ | logarithmic decrement; $\quad \Lambda = T\delta = T/\tau$ |
| $u, v$ | velocity; $\quad v = ds/dt$ |
| $\omega$ | angular velocity; $\quad \omega = d\varphi/dt$ |
| $a$ | acceleration; $\quad a = dv/dt$ |
| $\alpha$ | angular acceleration; $\quad \alpha = d\omega/dt$ |
| $g$ | local acceleration of free fall |
| $g_n$ | standard acceleration of free fall |
| $c$ | speed of light in vacuum |
| $\beta$ | fraction of the speed of light in vacuum; $\quad \beta = v/c$ |
| $(x_0, x_1, x_2, x_3)$ | relativistic coordinates; $\quad x_0 = ct$ |
| $(x_1, x_2, x_3, x_4)$ | $x_4 = ict$ |

[†] In molecular spectroscopy the wave number in vacuum $v/c$ is denoted by $\tilde{v}$.

**i** A more comprehensive list of these quantities with definitions and explanations can be found in ISO 80000-3 [9].

## 6.2 Mechanics

**Tab. 6.2:** Quantities of mechanics

| | |
|---|---|
| $m$ | mass |
| $\rho$ | mass density, density; $\rho = m/V$ |
| $d$ | relative density; $d = \rho/\rho_0$ |
| $v$ | specific volume; $v = V/m = 1/\rho$ |
| $\mu, m_r$ | reduced mass; $\mu = m_1 m_2/(m_1 + m_2)$ |
| $p$ | momentum; $p = mv$ |
| $L, J$ | angular momentum, spin; $L = r \times p$ |
| $I_a$ | axial moment of inertia; $I_{a,y} = \int x^2 \, dx \, dy$ |
| $I_p$ | polar moment of inertia; $I_p = \int (x^2 + y^2) \, dx \, dy$ |
| $I, J$ | moment of inertia; $I_z = \int (x^2 + y^2) \, dm$ |
| $F$ | force |
| $I$ | shock pulse; $I = \int F \, dt$ |
| $T$ | torsional moment (torque) |
| $G, (W, P)$ | weight |
| $M$ | torque |
| $H$ | torque impulse; $H = \int M \, dt$ |
| $p$ | pressure |
| $\sigma$ | normal stress |
| $\tau$ | shear stress |
| $G$ | gravitational constant; $F(r) = Gm_1 m_2/r^2$ |
| $\varepsilon$ | stretch, relative elongation; $\varepsilon = \Delta l/l_0$ |
| $E, (Y)$ | modulus of elasticity; $\sigma = E\varepsilon$ |
| $\gamma$ | shear |
| $G, \mu$ | shear modulus; $\tau = G\gamma$ |
| $\epsilon_{ij}$ | strain tensor |
| $\sigma_{ij}$ | stress tensor |
| $\theta$ | relative volume change; $\theta = \Delta V/V$ |
| $K$ | compression modulus; $p = -K\vartheta$ |

**Tab. 6.2:** continued

| | |
|---|---|
| $\mu, \nu$ | Poisson number |
| $\eta, (\mu)$ | dynamic viscosity |
| $\nu$ | kinematic viscosity;   $\nu = \eta/\rho$ |
| $\mu, (f)$ | friction coefficient |
| $\gamma, \sigma$ | surface tension, interfacial tension |
| $E, W$ | energy |
| $E_p, V, \Phi$ | potential energy |
| $E_k, T, K$ | kinetic energy |
| $W, A$ | work |
| $P$ | power |
| $\eta$ | efficiency |
| $H$ | Hamiltonian (function) |
| $W, S_p$ | Hamilton's principal function;   $W = \int L \mathrm{d}t$ |
| $L$ | Lagrangian (function) |

---

i  A more comprehensive list of these quantities with definitions and explanations can be found in ISO 80000-4 [10].

# 6.3 Statistical physics

**Tab. 6.3:** Quantities of statistical physics

| | |
|---|---|
| $N$ | number of particles |
| $n$ | number density of particles; $\quad n = N/V$ |
| $N_A, L$ | Avogadro constant |
| $m$ | molecular mass |
| $r, (x, y, z)$ | particle position vector and its components |
| $c, (c_x, c_y, c_z)$ | particle velocity vector and its components |
| $u, (u_x, u_y, u_z)$ | |
| $p, (p_x, p_y, p_z)$ | particle momentum vector and its components |
| $c_0, u_0, \langle c \rangle, \langle u \rangle$ | average velocity |
| $\bar{c}_0, \bar{u}_0, \langle c \rangle, \langle u \rangle$ | average speed |
| $\hat{c}, \hat{v}$ | most probable speed |
| $l, \lambda$ | mean free path |
| $\varepsilon$ | energy parameter of molecular attraction |
| $\varphi_{ij}, V_{ij}$ | interaction energy between particles $i$ and $j$ |
| $f(c)$ | velocity distribution function; $\quad \iiint f \, dc_x \, dc_y \, dc_z$ |
| $H$ | Boltzmann $H$ function |
| $q$ | generalised coordinate |
| $p$ | generalised momentum |
| $\Omega$ | volume in $\gamma$ phase space |
| $T$ | thermodynamic temperature |
| $k_B$ | Boltzmann constant |
| $\beta$ | $1/(k_B T)$ (as an argument of exponential functions) |
| $R$ | universal gas constant |
| $Q, Z$ | canonical partition function |
| $s$ | symmetry number |
| $D$ | diffusion coefficient |
| $D_{td}$ | thermal diffusion coefficient |
| $k_T$ | thermal diffusion ratio; $\quad k_T = D_T/D$ |

**Tab. 6.3:** continued

| | |
|---|---|
| $\alpha_T$ | thermal diffusion factor |
| $\Theta$ | characteristic temperature |
| $\Theta_D$ | Debye temperature;   $\Theta_D = h\nu_D/k_B$ |
| $\Theta_E$ | Einstein temperature;   $\Theta_E = h\nu_E/k_B$ |
| $\Theta_{rot}$ | rotational characteristic temperature;   $\Theta_r = h^2/(8\pi^2 I k_B)$ |
| $\Theta_{vib}$ | vibrational characteristic temperature;   $\Theta_v = h\nu/k_B$ |

A more comprehensive list of these quantities with definitions and explanations can be found in ISO 80000-5 [11].

## 6.4 Thermodynamics

If it is necessary, the index m is added to a symbol, in order to distinguish molar quantities from quantities referring to the whole system. For special quantities the corresponding lower case letters are used.

**Tab. 6.4:** Quantities of thermodynamics

| | |
|---|---|
| $Q$ | quantity of heat |
| $W, A$ | work |
| $T$ | thermodynamic temperature |
| $t, \vartheta$ | Celsius temperature [a] |
| $Z$ | partition function |
| $S$ | entropy |
| $U$ | internal energy |
| $A, F$ | free energy, Helmholtz function;  $F = U - TS$ |
| $H$ | enthalpy;  $H = U + pV$ |
| $G$ | free enthalpy, Gibbs function;  $G = H - TS$ |
| $J$ | Massieu function;  $J = -F/T$ |
| $Y$ | Planck function;  $Y = -G/T$ |
| $p$ | pressure |
| $\beta$ | pressure coefficient;  $\beta = (\partial p/\partial T)_V$ |
| $\alpha_p, \alpha$ | relative pressure coefficient;  $\alpha_p = (1/p)\,(\partial p/\partial T)_V$ |
| $\kappa_T, \kappa$ | compressibility;  $\kappa_T = -(1/V)\,(\partial V/\partial p)_T$ |
| $\alpha_l$ | linear expansion coefficient;  $\alpha_l = (1/l)\,(\partial l/\partial T)_p$ |
| $\alpha_V, \gamma$ | cubic expansion coefficient;  $\alpha_V = (1/V)\,(\partial V/\partial T)_p$ |
| $\lambda$ | thermal conductivity, heat conductivity |
| $c_p, c_V$ | specific heat capacity;  $c = C/m$ |
| $C_p, C_V$ | heat capacity |
| $\mu$ | Joule-Thomson coefficient |
| $\kappa$ | isentropic exponent; [b]  $\kappa = -(V/p)\,(\partial p/\partial V)_S$ |
| $\gamma, (\kappa)$ | ratio of specific heat capacities |

**Tab. 6.4:** continued

| | |
|---|---|
| $\Phi, (q)$ | heat flow rate |
| $q, (\varphi)$ | density of heat flow rate |
| $\kappa, k, K, (\lambda)$ | thermal conductivity |
| $a$ | thermal diffusivity; $\quad a = \lambda / \rho c_P$ |

[a] When symbols for both time as well as Celsius temperature are required, $t$ shall be used to denote time and $\vartheta$ to denote temperature.

[b] For an ideal gas, the isentropic exponent $\kappa$ is identical to the ratio of specific heat capacities $\gamma$.

---

i  A more comprehensive list of these quantities with definitions and explanations can be found in ISO 80000-5 [11].

# 6.5 Electricity and magnetism

**Tab. 6.5:** Quantities of electricity and magnetism

| | |
|---|---|
| $Q, q$ | (electric) charge, quantity of electricity |
| $\rho$ | charge density, density of volume charge |
| $\sigma$ | surface charge density |
| $\varphi, V$ | electric potential |
| $U, V$ | (electric) potential difference, voltage |
| $E$ | electric field (strength) |
| $D$ | electric displacement |
| $C$ | (electric) capacitance |
| $\varepsilon$ | permittivity; $\quad D = \varepsilon E$ |
| $\varepsilon_0$ | (absolute) permittivity of vacuum |
| $\varepsilon_r$ | relative permittivity; $\quad \varepsilon_r = \varepsilon/\varepsilon_0$ |
| $P$ | electric polarisation; $\quad D = \varepsilon_0 E + P$ |
| $\chi_e$ | electric susceptibility |
| $\alpha, \gamma$ | electric polarisability |
| $p$ | electric dipole moment |
| $I, (i)$ | (electric) current |
| $j, (J)$ | (electric) current density |
| $H$ | magnetic field (strength) |
| $B$ | magnetic induction, magnetic flux density |
| $\Phi$ | magnetic flux |
| $\mu$ | permeability; $\quad B = \mu H$ |
| $\mu_0$ | (absolute) permeability of vacuum |
| $\mu_r$ | relative permeability; $\quad \mu_r = \mu/\mu_0$ |
| $M$ | magnetisation; $\quad B = \mu_0 (H + M)$ |
| $\chi_m$ | magnetic susceptibility |
| $m, \mu_m$ | electromagnetic moment |
| $R$ | (electric) resistance |
| $X$ | reactance |

**Tab. 6.5:** continued

| | |
|---|---|
| $Q$ | quality factor;   $Q = \lvert X \rvert / R$ |
| $Z$ | impedance; [a]  $Z = R + \mathrm{i}X$ |
| $Y$ | admittance; [b]  $Y = 1/Z = G + \mathrm{i}B$ |
| $\rho$ | (electric) resistivity |
| $\sigma, \gamma$ | (electric) conductivity;   $\sigma = 1/\rho$ |
| $L$ | self-inductance |
| $M, L_{12}$ | mutual inductance |
| $k$ | coupling coefficient; [c]  $k = L_{12}/(L_1 L_2)^{1/2}$ |
| $\delta$ | loss angle |
| $w$ | electromagnetic energy density |
| $S$ | Poynting vector |
| $A$ | (magnetic) vector potential |

[a] $\lvert Z \rvert$ apparent impedance
[b] $\lvert Y \rvert$ apparent admittance
[c] according to DIN 1304:   $k = \lvert L_{12} \rvert / (L_1 L_2)^{1/2}$

---

ℹ A more comprehensive list of these quantities with definitions and explanations can be found in ISO 80000-6 [12].

## 6.6 Radiation and light

In several cases, the same symbol is used for a pair of corresponding radiometric and photometric quantities with the convention that subscripts e for *energetic* and v for *visible* are attached, otherwise there could be confusion between these quantities.

**Tab. 6.6:** Quantities of radiation and light

| | |
|---|---|
| $Q, (Q_e), W$ | radiant energy |
| $w$ | radiant energy density |
| $w_\lambda$ | spectral concentration of radiant energy density;   $w = \int w_\lambda d\lambda$ |
| $\Phi, (\Phi_e), P$ | radiant flux, radiant power |
| $\Phi_\lambda$ | spectral radiant flux;   $\Phi = \int \Phi_\lambda d\lambda$ |
| $\varphi$ | radiant flux density |
| $I, (I_e)$ | radiant intensity;   $\Phi = \int I \, d\Omega$ |
| $I_\lambda$ | spectral radiant intensity;   $I = \int I_\lambda d\lambda$ |
| $E, (E_e)$ | irradiance;   $\Phi = \int E \, dS$ |
| $L, (L_e)$ | radiance;   $I = \int L \cos \vartheta \, dS$ |
| $M, (M_e)$ | radiant exitance;   $\Phi = \int M \, dS$ |
| $\sigma$ | Stefan-Boltzmann constant;   $\sigma = 2\pi^5 k_B^4 / \left( 15 h^3 c^2 \right)$ |
| $c_1$ | first radiation constant;   $c_1 = 2\pi h c^2$ |
| $c_1$ | second radiation constant;   $c_2 = hc/k_B$ |
| $K$ | luminous efficacy;   $K = \Phi_v / \Phi_e$ |
| $K(\lambda)$ | spectral luminous efficacy |
| $K_m$ | maximum spectral luminous efficacy |
| $V(\lambda)$ | spectral luminous efficiency;   $V(\lambda) = K(\lambda)/K_m$ |
| $Q, (Q_v)$ | quantity of light |
| $\Phi, (\Phi_v)$ | luminous flux |
| $I, (I_v)$ | luminous intensity;   $\Phi = \int I d\Omega$ |
| $I_{v\lambda}$ | spectral luminous intensity;   $I = \int I_{v\lambda} d\lambda$ |
| $E, (E_v)$ | illuminance, illumination;   $\Phi = \int E dS$ |
| $L, (L_v)$ | luminance;   $I = \int L \cos \vartheta \, dS$ |

**Tab. 6.6:** continued

| | |
|---|---|
| $M, (M_v)$ | luminous exitance;  $\Phi = \int M \, dS$ |
| $\epsilon$ | emissivity |
| $\epsilon(\lambda)$ | spectral emissivity |
| $\alpha$ | absorptivity;  $\alpha = \Phi_a / \Phi_0$ |
| $\alpha(\lambda)$ | spectral absorptivity;  $\alpha(\lambda) = \Phi_{a\lambda} / \Phi_{0\lambda}$ |
| $\rho$ | reflectance;  $\rho = \Phi_r / \Phi_0$ |
| $\rho(\lambda)$ | spectral reflectance;  $\rho(\lambda) = \Phi_{r\lambda} / \Phi_{0\lambda}$ |
| $\tau$ | transmittance;  $\tau = \Phi_{tr} / \Phi_0$ |
| $\tau(\lambda)$ | spectral transmittance;  $\tau(\lambda) = \Phi_{tr\lambda} / \Phi_{0\lambda}$ |
| $\mu$ | linear attenuation coefficient |
| $a$ | linear absorption coefficient |
| $c_0$ | speed of light in vacuum |
| $n$ | refractive index, optical density;  $n = c_0 / c_M$ |

A more comprehensive list of these quantities with definitions and explanations can be found in ISO 80000-7 [13].

# 6.7 Acoustics

**Tab. 6.7:** Quantities of acoustics

| | |
|---|---|
| $c$ | velocity of sound |
| $c_l$ | velocity of longitudinal waves |
| $c_t$ | velocity of transverse waves |
| $c_g$ | group velocity |
| $P, P_a$ | sound power |
| $\rho$ | (sound) reflection coefficient;  $\rho = P_r/P_0$ |
| $\alpha_a, (\alpha)$ | (sound) absorption coefficient;  $\alpha_a = 1 - \rho$ |
| $\tau$ | (sound) transmission coefficient;  $\tau = P_{tr}/P_c$ |
| $\delta$ | (sound) dissipation factor;  $\delta = \alpha_a - \tau$ |
| $L_N$ | loudness level |
| $L_P, L_W$ | sound power level |
| $L_p$ | sound pressure level |

A more comprehensive list of these quantities with definitions and explanations can be found in ISO 80000-8 [6]. **i**

## 6.8 Quantum mechanics

**Tab. 6.8:** Quantities of quantum mechanics

| | |
|---|---|
| $\psi$ | wave function |
| $\psi^*$ | complex conjugate of the wave function |
| $P$ | probability density; $P \propto \psi^* \psi$ |
| $S$ | probability current density; $S = \dfrac{i\hbar}{2m} \left( \psi \nabla \psi^* - \psi^* \nabla \psi \right)$ |
| $\rho$ | charge density of electrons; $\rho = -eP$ |
| $j$ | current density of electrons; $j = -eS$ |
| $\langle \psi |$ | Dirac's bra vector |
| $| \psi \rangle$ | Dirac's ket vector |
| $\langle A \rangle, \overline{A}$ | expectation (value) of $A$ |
| $[A, B], [A, B]_-$ | commutator of $A$ and $B$; $[A, B] = AB - BA$ |
| $[A, B]_+$ | anticommutator of $A$ and $B$; $[A, B]_+ = AB + BA$ |
| $A_{ij}$ | matrix element; $\int \psi_i^* A \psi_j \, \mathrm{d}\tau$ |
| $A^\dagger$ | Hermitian conjugate of $A$ |
| $-i\hbar\nabla$ | momentum operator (in coordinate representation) |
| $a, b, \alpha, \beta$ | annihilation operators |
| $a^\dagger, b^\dagger, \alpha^\dagger, \beta^\dagger$ | creation operators |
| $\sigma, (\sigma_x, \sigma_y, \sigma_z)$ | Pauli matrices [a] |
| | $$\sigma_x = \begin{pmatrix} 0 & 1 \\ 1 & 0 \end{pmatrix}, \sigma_y = \begin{pmatrix} 0 & -i \\ i & 0 \end{pmatrix}, \sigma_z = \begin{pmatrix} 0 & 1 \\ 0 & -1 \end{pmatrix}$$ |
| $I$ | unit matrix; $I = \begin{pmatrix} 1 & 0 \\ 0 & 1 \end{pmatrix}$ |

**Tab. 6.8:** continued

| $\alpha, (\alpha_x, \alpha_y, \alpha_z, \beta)$ | Dirac $(4 \times 4)$ matrices [b,c] |
|---|---|

$$\alpha_x = \begin{pmatrix} 0 & \sigma_x \\ \sigma_x & 0 \end{pmatrix}, \alpha_y = \begin{pmatrix} 0 & \sigma_y \\ \sigma_y & 0 \end{pmatrix},$$

$$\alpha_z = \begin{pmatrix} 0 & \sigma_z \\ \sigma_z & 0 \end{pmatrix}, \beta = \begin{pmatrix} I & 0 \\ 0 & -I \end{pmatrix}$$

[a] The Pauli matrices can also be denoted by $(\sigma_1, \sigma_2, \sigma_3)$
[b] The Dirac matrices can also be denoted by $(\alpha_1, \alpha_2, \alpha_3, \alpha_4)$
[c] Occasionally, a different representation of the Dirac matrices is used.

## 6.9 Atomic and nuclear physics

**Tab. 6.9:** Quantities of atomic and nuclear physics

| | |
|---|---|
| $A$ | nucleon number, mass number |
| $Z$ | proton number, atomic number |
| $N$ | neutron number; $N = A - Z$ |
| $e$ | elementary charge (positive: proton, negative: electron) |
| $m_e$ | electron mass |
| $m_p$ | proton mass |
| $m_n$ | neutron mass |
| $m_\pi$ | meson mass |
| $m_N, m_N \left({}^A\mathrm{X}\right)$ | nuclear mass (of nucleus ${}^A\mathrm{X}$) |
| $m_a, m_a \left({}^A\mathrm{X}\right)$ | atomic mass (of nuclide ${}^A\mathrm{X}$) |
| $A_r$ | relative atomic mass |
| $h$ | Planck constant |
| $n, n_i$ | principal quantum number |
| $L, l_i$ | orbital angular momentum quantum number |
| $S, s_i$ | spin quantum number |
| $J, j_i$ | total angular momentum quantum number |
| $M, m_i$ | magnetic quantum number |
| $I, J$ | nuclear spin quantum number [†] |
| $F$ | hyperfine quantum number |
| $J, K$ | rotational quantum number |
| $v$ | vibrational quantum number |
| $Q$ | quadrupole moment |
| $R_\infty$ | Rydberg constant; $R_\infty = \mu_0^2 m_e e^4 c^3 / (8h^3)$ |
| $a_0$ | Bohr radius; $a_0 = \alpha / (4\pi R_\infty)$ |
| $\alpha$ | fine structure constant; $\alpha = \mu_0 c e^2 / (2h)$ |
| $\Delta$ | mass excess |
| $R$ | nuclear radius; $R = r_0 A^{1/3}$ |
| $\mu$ | magnetic moment of a particle |
| $\mu_p$ | magnetic moment of the proton |

**Tab. 6.9:** continued

| | |
|---|---|
| $\mu_n$ | magnetic moment of the neutron |
| $\mu_e$ | magnetic moment of the electron |
| $\mu_B$ | Bohr magneton;   $\mu_B = e\hbar/(2m_e)$ |
| $g$ | $g$-factor |
| $\gamma$ | gyromagnetic ratio |
| $\omega_L$ | Larmor circular frequency |
| $\Gamma$ | level width |
| $\tau, \tau_m$ | mean life |
| $Q$ | reaction energy, disintegration energy |
| $\sigma$ | cross section |
| $\Sigma$ | macroscopic cross section |
| $b$ | impact parameter |
| $\vartheta, \theta, \varphi$ | scattering angle |
| $\alpha$ | internal conversion coefficient |
| $Q$ | reaction energy, disintegration energy |
| $T_{1/2}$ | half life |
| $\lambda$ | decay constant |
| $A$ | activity (of a radioactive substance) |
| $\lambda_C$ | Compton wavelength;   $\lambda_C = h/(mc)$ |
| $r_e$ | electron radius |
| $\mu, \mu_l$ | linear attenuation coefficient |
| $\mu_a$ | atomic attenuation coefficient |
| $\mu_m$ | mass attenuation coefficient |
| $S, S_l$ | linear stopping power |
| $S_a$ | atomic stopping power |
| $R, S_l$ | linear range |
| $\alpha$ | recombination coefficient |

$^\dagger$ In atomic physics $I$ is used, in nuclear physics $J$.

A more comprehensive list of these quantities with definitions and explanations can be found in ISO 80000-10 [14].   **i**

## 6.10 Solid state physics

**Tab. 6.10:** Quantities of solid state physics

| | |
|---|---|
| $a, b, c$; $a_1, a_2, a_3$ | translation vectors of the crystal lattice |
| $R, R_0$ | lattice vector; $\quad R = n_1 a_1 + n_2 a_2 + n_3 a_3$ |
| $a^*, b^*, c^*$; $b_1, b_2, b_3$ | translation vectors of the reciprocal crystal lattice; |
| | $\quad a_i \cdot b_k = 2\pi \delta_{ik}$, ($\delta_{ik}$ denotes the Kronecker symbol)[a] |
| $G$ | reciprocal lattice vector; $\quad G \cdot R = 2\pi m, \; m \in \mathbb{Z}$ |
| $d$ | lattice plane spacing |
| $\vartheta$ | Bragg angle |
| $n$ | order of reflection |
| $\sigma$ | short range order parameter |
| $s$ | long range order parameter |
| $b$ | Burgers vector |
| $r, R$ | particle position vector[b] |
| $R_0$ | equilibrium position vector of an ion |
| $u$ | displacement vector of an ion |
| $Q_i$ | normal coordinates |
| $e$ | polarization vector |
| $D$ | Debye-Waller factor |
| $q_D$ | Debye (angular) wave number |
| $\omega_D$ | Debye (angular) frequency |
| $\Gamma, \gamma$ | Grüneisen parameter; $\quad \gamma = \alpha V / (\kappa C_v)$ |
| | ($\alpha$: cubic expansion coefficient, $\kappa$: compressibility) |
| $\alpha$ | Madelung constant |
| $l, l_e$ | mean free path of electrons |
| $\Lambda, l_{ph}$ | mean free path of phonons |
| $v_{dr}$ | drift velocity |
| $v_{th}$ | thermal velocity; $\quad v_{th} = \sqrt{\dfrac{3kT}{m}}$ |
| $\mu$ | mobility |

**Tab. 6.10:** continued

| | |
|---|---|
| $\psi_k(r)$ | one-electron wave function |
| $u_k(r)$ | Bloch wave function; $\psi_k(r) = u_k(r)\,e^{ik\cdot r}$ |
| $N_E, \rho$ | density of states; $dN(E)/dE = N_E(E) = \rho(E)$ |
| $g, N_\omega$ | (spectral) density of vibrational modes |
| $J$ | exchange integral |
| $\rho_{ik}$ | resistivity tensor |
| $\sigma_{ik}$ | electric conductivity tensor |
| $\lambda_{ik}$ | thermal conductivity tensor |
| $\rho_R$ | residual resistivity |
| $\tau$ | relaxation time; |
| $L$ | Lorenz coefficient, Lorenz number; $L = \lambda/(\sigma T)$ |
| $R_H, A_H$ | Hall coefficient |
| $A_E, P_E$ | Ettinghausen coefficient |
| $A_N$ | first Ettinghausen-Nernst coefficient |
| $A_{RL}, S_{RL}$ | first Righi-Leduc coefficient |
| $E_{ab}, \Theta_{ab}$ | thermoelectromotive force between substances a and b |
| $S_{ab}, \epsilon_{ab}$ | Seebeck coefficient for substances a and b; $S_{ab} = dE_{ab}/dT$ |
| $\Pi_{ab}$ | Peltier coefficient for substances a and b |
| $\mu, (\tau)$ | Thomson coefficient |
| $D$ | diffusion coefficient; $D = D_o \exp\left(\dfrac{E}{kt}\right)$ |
| $D_e$ | effective diffusion coefficient |
| $k$ | segregation coefficient |
| $\varphi, \Phi$ | work function; [c] $\Phi = e\varphi$ |
| $A$ | Richardson constant; $j = AT^2 \exp\left(-\Phi/kT\right)$ |
| $n, n_n, n_-$ | electron number density [d] |
| $p, n_p, n_+$ | hole number density [e] |
| $n_d$ | donor number density |
| $n_a$ | acceptor number density |
| $n_i$ | intrinsic number density |

**Tab. 6.10:** continued

| | |
|---|---|
| $E_g$ | energy gap; |
| $E_d$ | donor ionization energy |
| $E_a$ | acceptor ionization energy |
| $E_F, \epsilon_F$ | Fermi energy |
| $k$ | propagation vector (of particles) |
| $q$ | propagation vector (of phonons) |
| $k_F$ | Fermi propagation vector |
| $a$ | electron annihilation operator |
| $a^\dagger$ | electron creation operator |
| $b$ | phonon annihilation operator |
| $b^\dagger$ | phonon creation operator; |
| $m_n^*, m_p^*$ | effective mass of electrons or holes |
| $\mu_n, \mu_p$ | mobility of electrons or holes |
| $b$ | mobility ratio; $b = \mu_n/\mu_p$ |
| $D_n, D_p$ | diffusion coefficient of electrons or holes |
| $L_n, L_p$ | diffusion length of electrons or holes |
| $\tau_n, \tau_p$ | life time of electrons or holes |
| $\Theta, \Theta_w$ | characteristic temperature (Weiss temperature) |
| $T_C$ | Curie temperature |
| $T_N$ | Néel temperature |
| $T_c$ | superconductor critical transition temperature |
| $H_{c1}, H_{c2}, H_{c3}$ | superconductor critical field strength of a type II superconductor [f] |
| $H_c$ | superconductor (thermodynamic) critical field strength |
| $\Delta$ | superconductor energy gap |
| $\lambda_L$ | London penetration depth |
| $\xi$ | coherence length |
| $\kappa$ | Landau-Ginzburg parameter; $\kappa = \lambda_L \xi \sqrt{2}$ |
| $\Phi_0$ | flux quantum; [g] $\Phi_0 = h/(2e)$ |
| $h, k, l;$ | Miller indices [h] |
| $(h, k, l),$ | single plane or set of parallel planes in a lattice |

**Tab. 6.10:** continued

| | |
|---|---|
| $\{h, k, l\}$, | full set of planes in a lattice equivalent by symmetry |
| $[u, v, w]$ | direction in a lattice |
| $\langle u, v, w \rangle$ | full set of directions in a lattice equivalent by symmetry |

[a] In crystallography, however, $a_i \cdot b_k = \delta_{ik}$ applies.

[b] The notation $r$ is used for electrons and the notation $R$ is used for ions.

[c] The symbol $W$ is used for the quantity $W = \Phi + \mu$, where $\mu$ is the electrochemical potential, which at $T = 0$ is equal to the Fermi energy $E_F$.

[d] In general, the indices n or $-$ can be used to denote electrons.

[e] In general, the indices p or $+$ can be used to indicate holes.

[f] $H_{c1}$: for the magnetic flux entering the superconductor,
$H_{c2}$: for the disappearance of volume superconductivity,
$H_{c3}$: for the disappearance of surface superconductivity.

[g] $1/\Phi_0 = 2e/h$ is also called the characteristic constant for macroscopic coherence in superconductors.

[h] The Miller indices may also be denoted by $h_1, h_2, h_3$.

Note:   If the letter symbols in bracketed expressions are replaced by numbers, then it is usual to omit the commas and to indicate a negative numerical value by a crossbar above the number (instead of $[-1, 1, 0]$, e. g. $[\bar{1}10]$ is written).

A more comprehensive list of these quantities with definitions and explanations can be found in ISO 80000-12 [15].

## 6.11 Molecular spectroscopy

**Tab. 6.11:** Quantities of molecular spectroscopy

| | |
|---|---|
| $\Lambda, \lambda_i$ | quantum number of component electronic orbital angular momentum vector along the symmetry axis |
| $\Sigma, \sigma_i$ | quantum number of component of electronic spin along the symmetry axis |
| $\Omega, \omega_i$ | quantum number of total electronic angular momentum vector along the symmetry axis |
| $I$ | quantum number of nuclear spin |
| $v$ | quantum number of vibrational mode |
| $d$ | degeneracy of vibrational mode |
| $l$ | quantum number of vibrational angular momentum (LM) |
| $J$ | quantum number of total angular momentum (excluding nuclear spin) |
| $M, M_J$ | quantum number of component of $J$ in the direction of an external field |
| $M_S$ | quantum number of component of $S$ in the direction of an external field |
| $F$ | quantum number of total angular momentum (including nuclear spin) |
| $K$ | quantum number of component of angular momentum along the symmetry axis (LM; excluding electron and nuclear spin);   $K = |\Lambda + l|$ |
| $M_F$ | quantum number of component of $F$ in the direction of an external field |
| $M_I$ | quantum number of component of $I$ in the direction of an external field |
| $S$ | quantum number of electronic spin |
| $N$ | quantum number of total angular momentum (LM and STM; excluding electron and nuclear spin); [a] $J = N + S$ |

**Tab. 6.11:** continued

| | |
|---|---|
| $P$ | quantum number of component of angular momentum along the symmetry axis (LM and STM; excluding nuclear spin);[b] for LM: $P = \|K + \Sigma\|$, for STM: $P = \|K + \Sigma\|$ |
| $T_e$ | electronic term;[c] $T_e = E_e/(hc)$ |
| $G$ | vibrational term; $G = E_{\text{vibr}}/(hc)$ |
| $\sigma_e, x_e$ | coefficients in the expression for the vibrational term (for DM); $$G = \sigma_e\left(v + \frac{1}{2}\right) - x_e\left(v + \frac{1}{2}\right)^2$$ |
| $\sigma_j, x_{jk}$ | coefficients in the expression for the vibrational term (for PM); $$G = \sum_j \sigma_j\left(v_j + \frac{1}{2}d_j\right) - \frac{1}{2}\sum_{j,k} x_{jk}\left(v_j + \frac{1}{2}d_j\right)\left(v_k + \frac{1}{2}d_k\right)$$ |
| $F$ | rotational term; $F = E_{\text{rot}}/(hc)$ |
| $I_A, I_B, I_C$ | principal moments of inertia;[d] $I_A \leq I_B \leq I_C$ |
| $A, B, C$ | rotational constants $A = h/(8\pi^2 c I_A), B = h/(8\pi^2 c I_B), C = h/(8\pi^2 c I_C)$ |
| $T$ | total term; $T = T_e + G + F$ |

[a] Case of loosely coupled electron spin.
[b] Case of tightly coupled electron spin.
[c] All energies are taken with respect to the ground state as reference level.
[d] For diatomic molecules, use $I$ and $A = h/(8\pi^2 c I)$.

*Note*: The following abbreviations have been used in the table:

| | |
|---|---|
| LM | linear molecules |
| STM | rotationally symmetric molecules |
| DM | diatomic molecules |
| PM | polyatomic molecules |

For further details see the *Report on Notation for the Spectra of Polyatomic Molecules (Joint Commission for Spectroscopy of IAU and IUPAP, 1954,* J. chem. Phys. **23** (1955) S. 1997.

A more comprehensive list of these quantities with definitions and explanations can be found in ISO 80000-9 [16]. **i**

## 6.12 Chemical physics

**Tab. 6.12:** Quantities of chemical physics

| | |
|---|---|
| $A_r$ | relative atomic mass (of a nuclide or an element) |
| $M_r$ | relative molar mass of a substance |
| $n, v$ | amount of substance [†] |
| $N$ | number of entities |
| $C, n$ | number concentration; $\quad C = N/V$ |
| $N_A, (L)$ | Avogadro constant |
| $M_B$ | molar mass of the substance B |
| $c_B$ | concentration of the substance B; $\quad c_B = n_B/V$ |
| $x_B$ | molar fraction of the substance B |
| $w_B$ | mass fraction of the substance B |
| $p_B$ | partial pressure of the substance B |
| $\varphi_B$ | volume fraction of the substance B |
| $r$ | molar ratio of solution |
| $m$ | molality of solution |
| $s$ | solubility |
| $\mu_B$ | chemical potential of the substance B (referred to one particle) |
| $\lambda_B$ | absolute activity of the substance B; $\quad \lambda_B = \exp\left(\dfrac{\mu_B}{kT}\right)$ |
| $a_B$ | relative activity of the substance B |
| $z_B$ | reduced activity of the substance B; $\quad z_B = \dfrac{(2\pi mkT)^{3/2}}{h^3}\lambda_B$ |
| $\gamma_B, f_B$ | activity coefficient of the substance B |
| $\pi$ | osmotic pressure |
| $g, \varphi$ | osmotic coefficient |
| $v_B$ | stoichiometric number of the substance B in a chemical reaction |
| $A$ | affinity off a chemical reaction |
| $\xi_B$ | extent of reaction of the substance B; $\quad d\xi_B = dn_B/v_B$ |
| $z$ | charge number of an ion |
| $I$ | ionic strength |

**Tab. 6.12:** continued

| | |
|---|---|
| $K$ | equilibrium constant |
| $E_A$, $E_a$ | (Arrhenius) activation energy |

† $\nu$ may be used as an alternative symbol for amount of substance when $n$ is used for number density of particles.

A more comprehensive list of these quantities with definitions and explanations can be found in ISO 80000-9 [16].

## 6.13 Plasma physics

**Tab. 6.13:** Quantities of plasma physics

| | |
|---|---|
| $\varepsilon$ | energy of a particle |
| $E_d$, $E_d(X)$ | dissociation energy (e. g. of molecule X) |
| $E_{ea}$ | electron affinity |
| $E_i$ | ionization energy |
| $x$ | degree of ionization; |
| $z$ | charge number of ion (positive or negative) |
| $n_z$ | number density of ions of charge number $z$;[a] |
| $x_z$ | degree of ionization for charge number $z \geq 1$; $\quad x_z = n_z/(n_z + n_{z-1})$ |
| $T_n$ | neutral particle temperature (gas temperature) |
| $T_i$ | ion temperature |
| $T_e$ | electron temperature |
| $n_e$ | electron number density |
| $\omega_{pe}$ | electron plasma (circular) frequency; $\quad \omega_{pe}^2 = n_e e^2/(\varepsilon_0 m_e)$ |
| $\lambda_D$ | Debye length |
| $a$ | charge of particle |
| $\omega_{ce}$ | electron cyclotron (circular) frequency; $\quad \omega_{ce} = (e/m_e)B$ |
| $\omega_{ci}$ | ion cyclotron (circular) frequency; $\quad \omega_{ci} = (ze/m_i)B$ |
| $\mu$, $m_r$ | reduced mass; $\quad \mu = m_1 m_2/(m_1 + m_2)$ |
| $b$ | impact parameter |
| $l$, $\lambda$ | mean free path |
| $\nu_{coll}$, $\nu_c$ | collision frequency |
| $\tau_{coll}$, $\tau_c$ | mean time interval between collisions; $\quad \tau_{coll} = 1/\nu_{coll}$ |
| $\tau$ | relaxation time; $\quad \tau = 1/k$ |
| $\sigma$ | cross section; $\quad \sigma = 1/(ln)$ |
| $k$ | rate coefficient |
| $k_m$ | one-body rate coefficient; $\quad dn_A/dt = -k_m n_A$ |
| $k_b$ | binary rate coefficient, two-body rate coefficient |
| | (z. B. X + Y $\rightarrow$ XY + $h\nu$ ); $\quad dn_{XY}/dt = -k_b n_X n_Y$ |

**Tab. 6.13:** continued

| | |
|---|---|
| $k_t$ | ternary rate coefficient, three-body rate coefficient |
| | (z. B. $X + Y + M \rightarrow XY + M^*$);   $dn_{XY}/dt = -k_t n_M n_X n_Y$ |
| $s_e$ | (electron) ionization efficiency;   $s_e = (\rho_0/\rho)dN/dx$ |
| | ($dN$ denotes the number of ion pairs formed by |
| | an ionizing electron travelling through $dx$ |
| | in the plasma at gas density $\rho$; |
| | $\rho_0$: gas density at $p_0 = 133.322$ Pa and $T_0 = 273.15$ K) |
| $\alpha$ | Townsend (electron) ionization coefficient [b] |
| $\beta$ | Townsend (ion) ionization coefficient |
| $\gamma$ | secondary electron emission coefficient |
| $v_{dr}$ | drift velocity |
| $\mu$ | mobility;   $\mu = v_{dr}/E$ |
| $D_+, D_-$ | positive or negative ion diffusion coefficient |
| $D_e$ | electron diffusion coefficient |
| $D_a, D_{amb}$ | ambipolar (ion–electron) diffusion coefficient; |
| | $$D_a = (D_+ \mu_e + D_e \mu_+)/(\mu_+ + \mu_e)$$ |
| $L_D, \Lambda$ | characteristic diffusion length |
| $v_i$ | ionization frequency |
| $\alpha_i$ | ion–ion recombination coefficient;   $dn_-/dt = -\alpha_i n_- n_+$ |
| $\alpha_e$ | electron–ion recombination coefficient;   $dn_e/dt = -\alpha_e n_e n_+$ |
| $p$ | plasma pressure |
| $p_m$ | magnetic pressure;   $p_m = B^2/(2\mu)$ |
| $\beta$ | (magnetic) pressure ratio;   $\beta = p/p_m$ |
| $v_m, \eta_m$ | magnetic diffusivity;   $v_m = 1/(\mu\sigma)$ |
| $v_A$ | Alfvén speed;   $v_A = B/(\mu\rho)^{1/2}$ |

[a] If only singly charged ions need to be considered, $n_{-1}$, and $n_{+1}$ may be represented by $n_-$ and $n_+$.
[b] The same name is also used for the quantity $\eta = \alpha/E$, where $E$ denotes the electric field strength.

*Note*:   The symbols used in the definition equations denote:

| | |
|---|---|
| $\mu$ | magnetic permeability |
| $\sigma$ | electric conductivity |
| $\rho$ | mass density |

## 6.14 Characteristic numbers

Characteristic numbers (i. e. quantities of dimension number) are used to describe transport phenomena or material parameters in fluids.

---

 The following symbols for the characteristics numbers are recommended in the international standard ISO 80000-11 [17].

---

### Momentum transport

| | |
|---|---|
| $Re$ | Reynolds number;   $Re = vl/\nu$ |
| $Eu$ | Euler number;   $Eu = \Delta p/\rho v^2$ |
| $Fr$ | Froude number;   $Fr = v(lg)^{-1/2}$ |
| $Gr$ | Grashof number;   $Gr = l^3 g\gamma\Delta T/\nu^2$ |
| $We$ | Weber number;   $We = \rho v^2 l/\sigma$ |
| $Ma$ | Mach number;   $Ma = v/c$ |
| $Kn$ | Knudsen number;   $Kn = \lambda/l$ |
| $Sr$ | Strouhal number;   $Sr = lf/v$ |

where the symbols used in the definition equations denote:

| | |
|---|---|
| $l$ | characteristic length |
| $\Delta T$ | characteristic temperature difference |
| $\rho$ | (mass) density |
| $\eta$ | dynamic viscosity |
| $\nu$ | kinematic viscosity;   $\eta/\rho$ |
| $g$ | acceleration of free fall |
| $\lambda$ | mean free path |
| $c$ | velocity of sound |
| $v$ | characteristic speed |
| $\Delta p$ | pressure difference |
| $\sigma$ | surface tension |
| $\gamma$ | cubic expansion coefficient;   $\gamma = -(1/\rho)\left(\partial\rho/\partial T\right)_p$ |
| $f$ | characteristic frequency |

## Transport of heat

$Fo$  Fourier number;  $Fo = a\,\Delta t/l^2$
$Pe$  Peclet number;  $Pe = vl/a = Re \times Pr$
$Ra$  Rayleigh number;  $Ra = l^3 g\gamma\Delta T/(va)$
$Nu$  Nusselt number;  $Nu = hl/\lambda$
$St$  Stanton number;  $St = h/(\rho vc_p)$

where the symbols used in the definition equations denote:

$l$  characteristic length
$\Delta t$  characteristic time interval
$g$  acceleration of free fall
$\eta$  dynamic viscosity
$v$  kinematic viscosity;  $\eta/\rho$
$c_p$  specific heat capacity at constant pressure
$\lambda$  thermal conductivity
$h$  heat transfer coefficient
$v$  characteristic speed
$\Delta T$  characteristic temperature difference
$\rho$  (mass) density
$\gamma$  cubic expansion coefficient;  $\gamma = -(1/\rho)\,(\partial\rho/\partial T)_p$
$a$  thermal diffusivity;  $a = \lambda/(\rho c_p)$

## Material parameters of fluids

$Pr$  Prandtl number;  $Pr = v/a$
$Sc$  Schmidt number;  $Sc = v/D$
$Le$  Lewis number;  $Le = a/D$

where the symbols used in the definition equations denote:

$\rho$  (mass) density
$\eta$  dynamic viscosity
$v$  kinematic viscosity;  $v = \eta/\rho$
$D$  diffusion coefficient
$c_p$  specific heat capacity at constant pressure
$\lambda$  thermal conductivity
$a$  thermal diffusivity;  $a = \lambda/(\rho c_p)$

**Transport of matter in a binary mixture**

$Fo^*$     Fourier number for mass transfer;   $Fo^* = D\,\Delta t/l^2$
$Pe^*$     Péclet number for mass transfer;   $Pe^* = vl/D$
$Gr^*$     Grashof number for mass transfer;   $Gr^* = l^3 g\beta^*\,\Delta x/v^2$
$Nu^*$     Nusselt number for mass transfer;   $Nu^* = kl/(\rho D)$
$St^*$     Stanton number for mass transfer;   $St^* = k/(\rho v)$

where the symbols used in the definition equations denote:

$l$     characteristic length
$\Delta t$     characteristic time interval
$g$     acceleration of free fall
$\eta$     dynamic viscosity
$v$     kinematic viscosity;   $v = \eta/\rho$
$D$     diffusion coefficient
$k$     mass transfer coefficient
$v$     characteristic speed
$\Delta x$     characteristic difference of the amount of substance
$\rho$     (mass) density
$\beta^*$     relative change of (mass) density;   $-(1/\rho)\,(\partial\rho/\partial x)_{T,p}$

**Magnetohydrodynamics**

$Rm$     magnetic Reynolds number;   $Rm = v\mu\sigma l$
$Al$     Alfvén number;   $Al = v/v_A$
$Ha$     Hartmann number;   $Ha = Bl\sqrt{\sigma/(\rho v)}$
$Co, Co_2$     Cowling number (second Cowling number);   $Co = B^2/(\mu\rho v^2)$
$Co_1$     first Cowling number;   $Co_1 = B^2 l\sigma/(\rho v)$

where the symbols used in the definition equations denote:

$\rho$     (mass) density
$l$     characteristic length
$v$     characteristic speed
$\eta$     dynamic viscosity
$v$     kinematic viscosity;   $v = \eta/\rho$
$\mu$     magnetic permeability
$B$     magnetic flux density
$\sigma$     electric conductivity
$v_A$     Alfvén speed;   $v_A = B(\rho\mu)^{-1/2}$

# 7 Elements, nuclides and particles

1. *Symbols for chemical elements* shall be written with upright letters. No full stop shall be placed after the symbol.

    *Examples*:   Ca   C   H   He

2. The *nucleon number, baryon number* (*mass number*) of a nuclide is expressed by an upper left superscript attached to the symbol (e. g. $^{14}N$).

3. If necessary, a right superscript can also be used to indicate a ionization state (e. g. $Ca_2^+$, $PO_4^{3-}$) or an excited atomic state (e. g. $He^*$). However, a metastable nuclear state is often treated as a nuclide in its own right (e. g. either $^{118}Ag^m$ or $^{118m}Ag$).

4. Roman numerals in a right superscript position are used to indicate the oxidation number.

    *Examples*:   $Pb_2^{II}Pb^{IV}O_4$;   $K_6Mn^{IV}Mo_9O_{32}$

5. The spectrum of a $z$-fold ionized atom is indicated by a small Roman numeral, whose value corresponds to $z + 1$, on the line after the symbol for the chemical element and separated from it by a thin space.

    *Examples*:   H I     spectrum of neutral hydrogen
                    Ca II    spectrum of singly ionised calcium
                    Al III   spectrum of dual ionised aluminium

6. In nuclear physics, in order to avoid confusion with molecular compounds, a left subscript may be used to indicate the number of protons and a right subscript to indicate the number of neutrons in a nucleus (e. g. $^{235}_{92}U_{143}$). Although all these subscripts are redundant, they can often be useful. The right subscript is usually omitted and should only be given if the left subscript is also present.

    The right subscript position is also used to indicate the number of atoms of a nuclide in a molecule. (e. g. $^{14}N_2$).

7. The common names for particles used as projectiles or products in nuclear reactions are given in table 7.1. In addition to the symbols given in the table, HI is an accepted designation for a general heavy ion (where no possibility of ambiguity exists).

https://doi.org/10.1515/9783111344119-007

**Tab. 7.1:** Symbols for particles.

| nucleon | N | deuteron, $^2H^+$ | d | electron | e, β |
|---|---|---|---|---|---|
| proton, $^1H^+$ | p | triton, $^3H^+$ | t | muon | μ |
| neutron | n | Helion, $^3He^{2+}$ | h | neutrino | $\nu, \nu_e, \nu_\mu,$ |
| | | | | | $\nu_\tau,$ |
| | | α-particle | α | | |
| Λ-particle | Λ | | | photon | γ |
| Σ-particle | Σ | pion | π | | |
| Ξ-particle | Ξ | K-meson | K | | |
| Ω-particle | Ω | | | | |

The charge state of a particle can be indicated by adding the signs −, + or the symbol 0 as a superscript.

*Examples:*   $\pi^-; \pi^+; \pi^0,\quad e^-; e^+,\quad \beta^-; \beta^+.$

If no charge is indicated in connection with the symbols p and e, these symbols refer to the positive proton and the negative electron, respectively. A dash ( ̄) or the tilde ( ̃) above the symbol for a particle is used to indicate the corresponding antiparticle; the notation p̄ for the antiproton is preferable to p⁻, but both ē and e⁺ (or β̄ and β⁺) are commonly used for the positron.

The symbol e (*not italic*) to designate the electron must not be confused with the symbol *e* (*italic*) for the magnitude of the elementary charge.

**Tab. 7.2:** General grouping of stable particles.

| bosons | γ, W, Z |
|---|---|
| leptons | e, $\nu_e$, μ, $\nu_\mu$, τ, $\nu_\tau$ |
| quarks (q) | u, d, c, s, t, b |
| mesons (q$\bar{q}$) | |
| without *strangeness* ($S = 0$) | $\pi^+, \pi^0, \pi^-, \eta, D^+, D^0$ |
| with *strangeness* ($S = 1$) | $K^+, K^0, (K_L, K_S), F^+$ |
| baryons (qqq) | |
| ($S = 0$) | p, n, $\Lambda_c^+$ |
| ($S = -1$) | $\Lambda, \Sigma^+, \Sigma^0, \Sigma^-$ |
| ($S = -2$) | $\Xi^0, \Xi^-$ |
| ($S = -3$) | $\Omega^-$ |

**8.** Simply stating that "P" is the symbol for the P-particle is often not very informative. In addition, the documentation *Review of Particle Properties* issued by the *Particle Data Group* (Lawrence Berkeley Laboratory and CERN) should be consulted as a reference for this and similar topics. It is beyond the scope of this book to give detailed information on all the relationships between particles. Thus, table 7.2 shows only a general grouping of the particles that are stable under the strong nuclear force and can actually be called *particles* and not just *resonant states*. Each fermion has an associated antiparticle; bosons are their own antiparticles.

The names for quarks are the symbols themselves; the labels u (*up*), d (*down*), c (*charm*), s (*strange*), t (*top* or *truth*), and b (*bottom* or *beauty*) serve only as mnemonics for these symbols.

The mesons $D^+$, $D^0$ and $F^+$ as well as the *charm* baryon $\Lambda_c^+$ have the *charm* quantum number $C = +1$. The B mesons have the *bottomness* quantum number $B = +1$.

**9.** The relative atomic masses given in the following list of names and symbols of natural chemical elements are taken from the publication of the *Commission on Isotopic Abundances and Atomic Weights (CIAAW)* of the *International Union of Pure and Applied Chemistry (IUPAC)* issued in 2021.

**Tab. 7.3:** Names and symbols of the chemical elements.

| $Z$ | Name | Symbol | Relative atomic mass [†] |
|---|---|---|---|
| 1 | hydrogen | H | [1.007 84; 1.008 11] |
| 2 | helium | He | 4.002 602(2) |
| 3 | lithium | Li | [6.938; 6.997] |
| 4 | beryllium | Be | 9.012 1831(5) |
| 5 | boron | B | [10.806; 10.821] |
| 6 | carbon | C | [12.0096; 12.0116] |
| 7 | nitrogen | N | [14.00643; 14.007 28] |
| 8 | oxygen | O | [15.999 03; 15.999 77] |
| 9 | fluorine | F | 18.998 403 163(6) |
| 10 | neon | Ne | 20.1797(6) |
| 11 | sodium | Na | 22.989 769 28(2) |
| 12 | magnesium | Mg | [24.304; 24.307] |
| 13 | aluminium | Al | 26.981 5384(3) |

**Tab. 7.3:** continued

| $Z$ | Name | Symbol | Relative atomic mass [†] |
|----|------|--------|--------------------------|
| 14 | silicon | Si | [28.084; 28.086] |
| 15 | phosphorus | P | 30.973 761998(5) |
| 16 | sulphur | S | [32.059; 32.076] |
| 17 | chlorine | Cl | [35.446; 35.457] |
| 18 | argon | Ar | [39.792; 39.963] |
| 19 | potassium | K | 39.0983(1) |
| 20 | calcium | Ca | 40.078(4) |
| 21 | scandium | Sc | 44.955 908(5) |
| 22 | titanium | Ti | 47.867(1) |
| 23 | vanadium | V | 50.9415(1) |
| 24 | chromium | Cr | 51.9961(6) |
| 25 | manganese | Mn | 54.938 043(2) |
| 26 | iron | Fe | 55.845(2) |
| 27 | cobalt | Co | 58.933 194(3) |
| 28 | nickel | Ni | 58.6934(4) |
| 29 | copper | Cu | 63.546(3) |
| 30 | zinc | Zn | 65.38(2) |
| 31 | gallium | Ga | 69.723(1) |
| 32 | germanium | Ge | 72.630(8) |
| 33 | arsenic | As | 74.921 595(6) |
| 34 | selenium | Se | 78.971(8) |
| 35 | bromine | Br | [79.901; 79.907] |
| 36 | krypton | Kr | 83.798(2) |
| 37 | rubidium | Rb | 85.4678(3) |
| 38 | strontium | Sr | 87.62(1) |
| 39 | yttrium | Y | 88.905 84(1) |
| 40 | zirconium | Zr | 91.224(2) |
| 41 | niobium | Nb | 92.906 37(1) |
| 42 | molybdenum | Mo | 95.95(1) |
| 43 | technetium | Tc | |
| 44 | ruthenium | Ru | 101.07(2) |

**Tab. 7.3:** continued

| $Z$ | Name | Symbol | Relative atomic mass [†] |
|---|---|---|---|
| 45 | rhodium | Rh | 102.905 49(2) |
| 46 | palladium | Pd | 106.42(1) |
| 47 | silver | Ag | 107.8682(2) |
| 48 | cadmium | Cd | 112.414(4) |
| 49 | indium | In | 114.818(1) |
| 50 | tin | Sn | 118.710(7) |
| 51 | antimony | Sb | 121.760(1) |
| 52 | tellurium | Te | 127.60(3) |
| 53 | iodine | I | 126.904 47(3) |
| 54 | xenon | Xe | 131.293(6) |
| 55 | caesium | Cs | 132.905 451 96(6) |
| 56 | barium | Ba | 137.327(7) |
| 57 | lanthanum | La | 138.905 47(7) |
| 58 | cerium | Ce | 140.116(1) |
| 59 | praseodymium | Pr | 140.907 66(1) |
| 60 | neodymium | Nd | 144.242(3) |
| 61 | promethium | Pm | |
| 62 | samarium | Sm | 150.36(2) |
| 63 | europium | Eu | 151.964(1) |
| 64 | gadolinium | Gd | 157.25(3) |
| 65 | terbium | Tb | 158.925 354(8) |
| 66 | dysprosium | Dy | 162.500(1) |
| 67 | holmium | Ho | 164.930 328(7) |
| 68 | erbium | Er | 167.259(3) |
| 69 | thulium | Tm | 168.934 218(6) |
| 70 | ytterbium | Yb | 173.045(10) |
| 71 | lutetium | Lu | 174.9668(1) |
| 72 | hafnium | Hf | 178.486(6) |
| 73 | tantalum | Ta | 180.947 88(2) |
| 74 | tungsten | W | 183.84(1) |
| 75 | rhenium | Re | 186.207(1) |

**Tab. 7.3:** continued

| $Z$ | Name | Symbol | Relative atomic mass [†] |
|---|---|---|---|
| 76 | osmium | Os | 190.23(3) |
| 77 | iridium | Ir | 192.217(2) |
| 78 | platinum | Pt | 195.084(9) |
| 79 | gold | Au | 196.966 570(4) |
| 80 | mercury | Hg | 200.592(3) |
| 81 | thallium | Tl | [204.382; 204.385] |
| 82 | lead | Pb | [206.14; 207.94] |
| 83 | bismuth | Bi | 208.980 40(1) |
| 84 | polonium | Po | |
| 85 | astatine | At | |
| 86 | radon | Rn | |
| 87 | francium | Fr | |
| 88 | radium | Ra | |
| 89 | actinium | Ac | |
| 90 | thorium | Th | 232.0377(4) |
| 91 | protactinium | Pa | 231.035 88(1) |
| 92 | uranium | U | 238.028 91(3) |
| 93 | neptunium | Np | |
| 94 | plutonium | Pu | |
| 95 | americium | Am | |
| 96 | curium | Cm | |
| 97 | berkelium | Bk | |
| 98 | californium | Cf | |
| 99 | einsteinium | Es | |
| 100 | fermium | Fm | |
| 101 | mendelevium | Md | |
| 102 | nobelium | No | |
| 103 | lawrencium | Lr | |
| 104 | rutherfordium | Rf | |
| 105 | dubnium | Db | |
| 106 | seaborgium | Sg | |

**Tab. 7.3:** continued

| $Z$ | Name | Symbol | Relative atomic mass [†] |
|-----|------|--------|--------------------------|
| 107 | bohrium | Bh | |
| 108 | hassium | Hs | |
| 109 | meitnerium | Mt | |
| 110 | darmstadtium | Ds | |
| 111 | roentgenium | Rg | |
| 112 | copernicium | Cn | |
| 113 | nihonium | Nh | |
| 114 | flerovium | Fl | |
| 115 | moscovium | Mc | |
| 116 | livermorium | Lv | |
| 117 | tennessine | Ts | |
| 118 | oganesson | Og | |

[†] The uncertainties of the relative atomic masses are given either by an interval or by a bracketed number. For example, the relative atomic mass of argon, [39.792; 39.963], means that its value is between 39.792 and 39.963. The relative atomic mass of iridium, 192.217(2), means that its value for normal materials lies between 192.215 and 192.219. For more information on the interpretation of the uncertainty please consult the recent *IUPAC Technical Report* by Possolo et al.

*Source*: Standard atomic weights of the elements 2021 (IUPAC Technical Report)
https://www.ciaaw.org/atomic-weights.htm
https://www.degruyter.com/document/doi/10.1515/pac-2019-0603/html

# 8 Quantum States

## 8.1 General rules

A letter symbol to denote the quantum state *of a system* shall be written as *uppercase upright letter*. A letter symbol to denote the quantum state *of a single particle* shall be written as *lowercase upright letter*.

## 8.2 Atomic spectroscopy

The letter symbols used to indicate the orbital angular momentum quantum number are

| $l =$ | 0 | 1 | 2 | 3 | 4 | 5 | 6 | 7 | 8 | 9 | 10 | 11 | ... |
|---|---|---|---|---|---|---|---|---|---|---|---|---|---|
| symbol | s | p | d | f | g | h | i | k | l | m | n | o | ... |
| $L =$ | 0 | 1 | 2 | 3 | 4 | 5 | 6 | 7 | 8 | 9 | 10 | 11 | ... |
| symbol | S | P | D | F | G | H | I | K | L | M | N | O | ... |

A right subscript denotes the total angular momentum quantum number $J$ or $j$. A left superscript denotes the spin multiplicity $2s + 1$ or $2S + 1$.

*Examples:*    $^3$D term        (spin multiplicity 3)

$^3$D$_2$ level        ($J = 2$)

d$_{3/2}$ electron    ($j = 3/2$)

The electron configuration of an atom is symbolically represented by

$$(nl)^k (n'l')^{k'} \dots ,$$

where $k$, $k'$, ... denote the number of electrons with principal quantum numbers $n$, $n'$, ... and orbital angular momentum quantum numbers $l$, $l'$, ... respectively. Instead of $l = 0, 1, 2, 3, \dots$, the quantum number symbols s, p, d, f, ... are usually used and the parentheses are omitted.

*Example:*    electron configuration of an atom:    $1s^2 2s^2 2p^3$ .

An atomic quantum state is specified by its quantum numbers. In the case of Russell-Saunders coupling (*LS* coupling), an *atomic term* is specified by $L$ and $S$ and an *atomic level* is specified by $L$, $S$ and $J$. An *atomic state* is specified by $L$, $S$, $J$ and $M_J$ or by $L$, $S$, $M_S$ and $M_L$.

https://doi.org/10.1515/9783111344119-008

## 8.3 Molecular spectroscopy

In contrast to the line spectra of atomic spectroscopy, molecular spectra are so-called *band spectra*. They consist of closely spaced individual lines that form overlapping groups ("bands"). The reason is that molecules, unlike atoms, can absorb or emit energy not only by means of electron transitions, but also when the atoms vibrate against each other and when the molecule rotates.

For *linear molecules* the letter symbols denoting the quantum number of the orbital angular momentum along the molecular axis are

$$
\begin{array}{lcccc}
\lambda = & 0 & 1 & 2 & \ldots \\
\text{symbol} & \sigma & \pi & \delta & \ldots \\
\Lambda = & 0 & 1 & 2 & \ldots \\
\text{symbol} & \Sigma & \Pi & \Delta & \ldots
\end{array}
$$

A left superscript indicates the spin multiplicity. For molecules with a symmetry centre, the parity symbol g (*gerade,* i. e. *even*) or u (*ungerade,* i. e. *odd*), indicating symmetric or antisymmetric behaviour on inversion, respectively, is attached as a right subscript. A + or − sign attached as a right superscript indicates the symmetry with respect to a reflection in any plane through the symmetry axis of the molecule.

*Examples:*   $\Sigma_g^+$,   $\Pi_u$,   $^2\Sigma$,   $^3\Pi$,   etc.

In case of linear molecules the letter symbols denoting the states of different angular momentum quantum numbers $l$ of degenerate vibrations are

$$
\begin{array}{lccccc}
l = & 0 & 1 & 2 & 3 & \ldots \\
\text{symbol} & \Sigma & \Pi & \Delta & \Phi & \ldots
\end{array}
$$

## 8.4 Nuclear spectroscopy

Spin and parity of a core state are denoted by $J^\pi$, where $\pi$ denotes the parity symbol (a plus sign for even parity and a minus sign for odd parity, respectively).

*Examples:*   $3^+$,   $2^-$,   etc.

A shell model configuration is symbolically represented by

$$
\nu(nl_j)^k (n'l'_{j'})^{k'} \ldots \pi(n''l''_{j''})^{k''} (n'''l'''_{j'''})^{k'''} ,
$$

where the symbol $\nu$ in front of the first bracket expression refers to the neutron shell and the symbol $\pi$ in front of the second one refers to the proton shell. Negative values of the superscripts indicate unoccupied sites in a complete shell. Instead of $l = 0, 1, 2, 3, \ldots$, the quantum number symbols s, p, d, f, … are also used (except for the value $l = 7$, which is denoted by $k$ for atoms and $j$ for nuclei).

*Example:* nuclear configuration $\nu(2d_{5/2})^6\pi(2p_{1/2})^2(1g_{9/2})^3$.

If neutrons and protons are in the same shell with a well-defined isospin $T$, then the notation $(nl_j)^\alpha$ is used, where $\alpha$ indicates the total number of nucleons.

*Example:* $(1f_{7/2})^5$.

## 8.5 Spectroscopic transitions

The upper and lower levels of a spectroscopic transition are denoted by $'$ and $''$.

*Examples:* $h\nu = E' - E''$, $\sigma = T' - T''$.

The designation of spectroscopic transitions is not uniform. In atomic spectroscopy[33] it is customary to write the lower state first and the upper state second; in molecular and polyatomic spectroscopy[34], however, the convention is reversed and one writes the upper state first and the lower state second. In both cases, the two state designations are connected by a dash (−).

*Examples:* $^2P_{1/2} - {}^2S_{1/2}$      for an electron transition
         $(J', K') - (J'', K'')$    for a rotational transition
         $\nu' - \nu''$          for a vibrational transition.

If it should be necessary to indicate whether the transition is an absorption or an emission process, the two state designations are connected by arrows ($\leftarrow$ and $\rightarrow$, respectively).

*Examples:* $^2P_{1/2} \rightarrow {}^2S_{1/2}$      emission to $^2S_{1/2}$
         $(J', K') \leftarrow (J'', K'')$    absorption from $(J'', K'')$

---

**33** see R. D. Cowan, *The Theory of Atomic Structure and Spectra* (Univ. of California Press, 1981),
**34** see *Report on Notation for the Spectra of Polyatomic Molecules*, J. Chem. Phys. **23** (1955) 1997.

If there is a risk of misunderstanding, the underlying convention for the order of states should be explicitly stated.

The difference between two quantum numbers is that of the upper state minus that of the lower state.

*Example:*   $\Delta J = J' - J''$

The branches of a rotation-vibration band are designated as follows

| $\Delta J$ | Branch |
|:---:|:---|
| −2 | O branch |
| −1 | P branch |
| 0 | Q branch |
| +1 | R branch |
| +2 | S branch |

## 8.6 Nomenclature in nuclear physics

### Nuclides

Atoms that are identical regarding atomic number (proton number) and mass number (nucleon number) shall be designated by the term *nuclide*, not by the term *isotope*. Different nuclides with the same mass number are called *isobaric nuclides* or *isobars*.

Different nuclides with the same atomic number are called *isotopes*. Since nuclides with the same number of protons are isotopes, nuclides with the same number of neutrons are sometimes called *isotones*.

The symbolic expression representing a nuclear reaction should correspond to the following scheme:

$$A\,(b,c)\,D\,,$$

where A denotes the nuclide in the initial state, D denotes the nuclide in the final state, b denotes the incident particle or photon, and c denotes the emitted particles (one or more).

*Examples:* $^{14}N\,(\alpha, p)\,^{17}O$ $^{59}Co\,(n, \gamma)\,^{60}Co$

$^{23}Na\,(\gamma, 3n)\,^{20}Na$ $^{31}P\,(\gamma, pn)\,^{29}Si$

## Characterization of interactions

Multipolarity of a transition:

| | |
|---|---|
| electric or magnetic monopole | E0 or M0 |
| electric or magnetic dipole | E1 or M1 |
| electric or magnetic quadrupole | E2 or M2 |
| electric or magnetic octupole | E3 or M3 |
| electric or magnetic $2^n$-pole | E$n$ or M$n$ |

Designation of the parity change during a transition:

| | |
|---|---|
| transition with parity change: | (yes) |
| transition without parity change: | (no) |

Notation for the covariant character of the coupling:

| | |
|---|---|
| S | scalar coupling |
| P | pseudo-scalar coupling |
| V | vector coupling |
| A | vector coupling (axial vector) |
| T | tensor coupling |

## Polarisation conventions

*Sign of polarization vector (Basel convention):*

The sign (orientation) of the polarization vector is chosen so that, in the case of a nuclear interaction, the positive polarization direction for particles with spin 1/2 coincides with the direction of the vector product

$$k_i \times k_o,$$

where $k_i$ and $k_o$, respectively, denote the wave vectors of the incident and emitted particles.

*Description of polarization effects (Madison convention)*:

In the symbolic expression for a nuclear reaction A (b,c) D, an arrow above a symbol indicates a particle that is initially in a polarized state or whose state of polarization is measured.

*Examples*:     A ($\vec{b}$,c) D     polarized incident particle

A ($\vec{b}$, $\vec{c}$) D     polarized incident particle;

the polarization of the emitted particle c

is measured (polarization transfer)

A (b,$\vec{c}$) D     unpolarized incident particle;

the polarization of the emitted particle c

is measured

$\vec{A}$ (b,c) D     unpolarized particle incident on

a polarized nuclide

$\vec{A}$ (b,$\vec{c}$) D     unpolarized particle incident on

a polarized nuclide; the polarization of the

emitted particle c is measured

A ($\vec{b}$,c) $\vec{D}$     polarized incident particle;

the polarization of the nuclide D in the

final state is measured

# A Conversions to US Customary Units

The United States of America were one of the 17 nations which signed the *Metre Convention* (French: *Convention du Mètre*) in Paris on 20 May 1875. However, for commercial and everyday use, the US Customary System is still used in the United States (US). Therefore, it is advisable to know the conversions between US Customary Units and SI Units.

---

**i** In September 1999 the NASA *Mars Climate Orbiter* failed due to a measurement-units-related software bug. The primary cause was that one piece of ground software supplied by Lockheed Martin produced results in a US Customary Unit, contrary to its Software Interface Specification (SIS), while a second system, supplied by NASA, expected those results to be in SI Units, in accordance with the SIS. Specifically, the software that calculated the total impulse produced by thruster firings produced results in lbf s (pound-force seconds). The trajectory calculation software subsequently used these results — expected to be in N s (newton seconds), i. e. incorrect by a factor of 4.45 — to update the predicted position of the spacecraft.[18]

---

The following tables summarize most commonly used US Customary Units and their (approximate) SI Unit equivalents.

**Tab. A.1:** Units of length

| US Unit | SI Unit equivalent |
| --- | --- |
| 1 in (inch) | 0.0254 m |
| 1 ft (foot) | 0.3048 m |
| 1 yd (yard) | 0.9144 m |
| 1 mi (statue mile) | 1609.344 m |

**Tab. A.2:** Units of area

| US Unit | SI Unit equivalent |
| --- | --- |
| 1 sq in (square inch) | $0.000\,645\,16\ \mathrm{m}^2$ |
| 1 sq ft (square foot) | $0.092\,903\,04\ \mathrm{m}^2$ |
| 1 sq yd (square yard) | $0.836\,127\,36\ \mathrm{m}^2$ |
| 1 acre | $4046.856\,4224\ \mathrm{m}^2$ |

https://doi.org/10.1515/9783111344119-009

**Tab. A.3:** Units of volume

| US Unit | SI Unit equivalent |
| --- | --- |
| 1 cu in (cubic inch) | 16.387 064 mL |
| 1 cu ft (cubic foot) | 28.316 846 592 L |
| 1 cu yd (cubic yard) | 0.764 554 857 984 m$^3$ |
| 1 gal (gallon) | 3.785 411 784 L |

**Tab. A.4:** Units of mass (weight) [†]

| US Unit | SI Unit equivalent |
| --- | --- |
| 1 gr (grain) | 64.798 91 mg |
| 1 oz (ounce) | 28.3495 g |
| 1 lb (pound) | 453.592 37 g |
| 1 ton (long ton) | 1016.046 9088 kg |

[†] The words *mass* and *weight* are interchangeably used in the USA.

**Tab. A.5:** Units of pressure

| US Unit | SI Unit equivalent |
| --- | --- |
| 1 psf (pound-force per square foot) | 47.880 258 980 336 Pa |
| 1 psi (pound-force per square inch) | 6894.757 293 168 370 Pa |
| 1 pdl/ft$^2$ (poundal per square foot) | 1.488 163 943 570 Pa |
| 1 atm (atmosphere) | 101 325 Pa |
| 1 in Hg (inch of mercury) [†] | 3386.388 157 894 730 Pa |
| 1 in H$_2$O (inch of water) [†] | 248.84 Pa |
| 1 ft H$_2$O (foot of water) [†] | 2986.080 Pa |

[†] at 60 °F (15.56 °C)

**Tab. A.6:** Units of energy and work

| US Unit | SI Unit equivalent |
| --- | --- |
| 1 Btu (British thermal unit; international) [38] | 1055.056 J |
| 1 Btu (British thermal unit; thermochemical) | 1054.350 J |
| 1 Btu (British thermal unit; mean) | 1055.87 J |
| 1 thm (therm) | 105 480 400 J |
| 1 Wh (watt-hour) | 3600.000 J |
| 1 ft lbf (foot-pound-force) | 1.355 817 948 331 4004 J |
| 1 ft pdl (foot-poundal) | 0.042 140 110 093 8048 J |

**Tab. A.7:** Units of power

| US Unit | SI Unit equivalent |
| --- | --- |
| 1 Btu/h (international British thermal unit per hour) | 0.293 07 W |
| 1 hp (mechanical horsepower) [42] | 745.69 W |
| 1 hp (boiler horsepower) | 9809.50 W |
| 1 ft lbf/s (foot-pound-force per second) | 1.3558 W |

Temperatures can be converted from the Fahrenheit scale to the Celsius scale, and vice versa, by using the equations

$$T_{\circ F} = \frac{9}{5} \times T_{\circ C} + 32 . \qquad T_{\circ C} = \frac{5}{9} \times (T_{\circ F} - 32) ,$$

# Bibliography

[1]    M. Krystek. "The term dimension in the international system of units". In: *Metrologia* 52 (2015), pp. 297–300.

[2]    SI Brochure. *The International System of Units (SI)*. 9th ed. Paris: BIPM, 2019.

[3]    IEC 80000-13. *Quantities and units — Part 13: Information science and technology*. International Organization for Standardization (ISO), 2008.

[4]    ISO/IEC 2382. *Information technology — Vocabulary*. Geneva: International Organization for Standardization (ISO), 2015.

[5]    IEC/CD 80000-15. *Quantities and units — Part 15: Logarithmic and related quantities*. International Organization for Standardization (ISO), 2021.

[6]    ISO 80000-8. *Quantities and units — Part 8: Acoustics*. International Organization for Standardization (ISO), 2020.

[7]    ISO 80000-1. *Quantities and units — Part 1: General*. International Organization for Standardization (ISO), 2009.

[8]    ISO 80000-2. *Quantities and units — Part 2: Mathematics*. International Organization for Standardization (ISO), 2019.

[9]    ISO 80000-3. *Quantities and units — Part 3: Space and time*. International Organization for Standardization (ISO), 2019.

[10]   ISO 80000-4. *Quantities and units — Part 4: Mechanics*. International Organization for Standardization (ISO), 2019.

[11]   ISO 80000-5. *Quantities and units — Part 5: Thermodynamics*. International Organization for Standardization (ISO), 2019.

[12]   IEC 80000-6. *Quantities and units — Part 6: Electromagnetism*. International Organization for Standardization (ISO), 2008.

[13]   ISO 80000-7. *Quantities and units — Part 7: Light and radiation*. International Organization for Standardization (ISO), 2019.

[14]   ISO 80000-10. *Quantities and units — Part 10: Atomic and nuclear physics*. International Organization for Standardization (ISO), 2019.

[15]   ISO 80000-12. *Quantities and units — Part 12: Condensed matter physics*. International Organization for Standardization (ISO), 2019.

[16]   ISO 80000-9. *Quantities and units — Part 9: Physical chemistry and molecular physics*. International Organization for Standardization (ISO), 2019.

[17]   ISO 80000-11. *Quantities and units — Part 11: Characteristic numbers*. International Organization for Standardization (ISO), 2019.

[18]   J. Oberg. "Why the Mars Probe went off course". In: *IEEE Spectrum* (1999).

https://doi.org/10.1515/9783111344119-010

# Index

https://doi.org/10.1515/9783111344119-011